Naturoids

On the Nature of the Artificial

Massimo Negrotti

University of Urbino, Italy

Naturoids

On the Nature of the Artificial

World Scientific

New Jersey • London • Singapore • Hong Kong

Published by

World Scientific Publishing Co. Pte. Ltd.

P O Box 128, Farrer Road, Singapore 912805

USA office: Suite 1B, 1060 Main Street, River Edge, NJ 07661

UK office: 57 Shelton Street, Covent Garden, London WC2H 9HE

British Library Cataloguing-in-Publication Data
A catalogue record for this book is available from the British Library.

ISBN 981-02-4932-2

This book is printed on acid-free paper.

Printed in Singapore by Uto-Print

Table of Contents

Part I: Theory

Part II: The Reality of the Artificial

1 Theory

1 The Icarus Syndrome

When Icarus, according to classic mythology, came crashing down near Samo because the wax that fixed his wings had melted, he probably repented bitterly for not having followed the advice of his father, Daedalus, who had warned him not to get too close to Sun. While a bird, in a similar circumstance, might only have been singed and would soon have descended to more reasonable heights, the artificial wings of Icarus could not withstand the test and he died suddenly.

On the other hand, we know that Homer, in the Iliad, describes the god Efesto, who creates the first woman, Pandora, from clay; Plato speaks of the mobile statue built by Daedalus himself; Argonauts, who searched for the Golden Fleece, had an artificial watch-dog at their disposal.

Attempts to reproduce natural instances dates back to antiquity. As it has been noted (Boas, 1927, Vogel, 2000), such attempts are even found in the early forms of art and in several fantastic poetic or religious accounts. As stated by de Solla Price

'Our story, then, begins with the deep-rooted urge of man to simulate the world about him through the graphic and plastic arts. The almost magical, naturalistic rock paintings of prehistoric caves, the ancient grotesque figurines and other "idols" found in burials, testify to the ancient origin of this urge in primitive religion' (de Solla Price, 1964).

Furthermore, automata – reproductions of human beings very different from one another in terms of their substance – have abounded throughout the history of human imagination, starting with the Bible through Faust and the R.U.R. (*Rezon's Universal Robot*) by Capek.

It would surely be very easy to expose, case by case, the weak points of each of the 'machines' quoted above, but there is another much more compelling question. What kind of relationship exists between these attempts to imitate nature and the technology of that particular historical period? In other words, is technology intrinsically intended to reproduce something existing in nature or is it developed with other aims in mind as well? After all, it is like when we ask ourselves: has man created technology only to reproduce nature?

The history of technology clearly shows that man, in designing and building objects, machines or processes is often motivated by a desire to imitate, but, in many other cases, he aims to control and dominate natural events, rather than reproduce them by means of expedients which are refined to varying degrees. As the mathematician Henry de Monantheuil stated in the xvi century,

'(...) man, being God's image, was invited to imitate him as a mechanician and to produce objects which could cope with those made by nature' (de Monantheuil, 1517, in Bredekamp, 1993).

In order to control natural events one needs to know them and, our technologies will vary in their effectiveness according to the accuracy of this knowledge. However, these technologies will not necessarily be designed to imitate the phenomena in question, but to adapt to nature in order to exploit its features and reach some useful goal.

Thus, the knowledge of some physical laws allowed us to build machines, like electrical or internal combustion engines, which enhance our ability to move physically. The advent of writing led to the invention of various writing tools which, today, are highly advanced thanks to computer technology. The knowledge and the exploitation of other natural laws made it possible to conceive, and then construct, systems which, like the cathode-ray tube, make the display of graphs or images, etc. possible.

In all these examples, and in the many more one could easily cite, there is no attempt to imitate, but, rather, an attempt to invent, to make behaviors, effects and events possible which, on the basis of our pure and simple natural condition, would not be attemptable.

To sum up, close to Icarus – and to all his descendants who constitute the world of the *artificialists* –the figure of Prometheus stands out. In giving fire to man, he awakened man's ability to invent, i. e., his ability to establish construction targets for objects or processes, as it were, additional and therefore different from those already existing in nature.

As we know, in doing so, unfortunately man created dangers and disasters. Prometheus himself was the first to pay with a tremendous torture for having taken possession of fire, which was a prerogative of the gods. Nevertheless, imitation and invention are two distinct circumstances and actions, and they require analyses which are likewise different.

In the following pages, we shall investigate and outline some fundamental aspects of the first of the above technologies, namely the *technology of the artificial* (or *technology of*

naturoids). A clear distinction is drawn between technology of
the artificial and *conventional technology*. Unlike the latter, the
former implicitly or explicitly, aims to reproduce something existing in nature. This distinction may be applied to all technological traditions, from the most pre-scientific to the most advanced ones, including nanotechnology, rightly defined by
Crandall and Lewis as

'a descriptive term for a particular state of our species' control
of materiality' (Crandall and Lewis,1997).

Nanotechnology, in fact, can be oriented either to reproduce natural things or processes, exhibiting different features,
or to produce new objects or materials.

The dichotomy between conventional and artificial technology tracks a distinction that, though accepted as a fact, has
never been clearly drawn, and it is very useful in trying to understand rationally a wide range of phenomena which are not
just technological.

We may consider, in fact, imitation and invention – i. e.
the basic human qualities which generate the technology of
the artificial and conventional technology – as human aptitudes which, on a social and cultural level, give rise to very different classes of behaviors and activities. Imitation, for instance, exhibits a variety of expressions which range from the
socialization of children to the fashion phenomena, and the
spreading of cultural models (scientific, technological, ethical,
religious, juridical, etc.).

Invention, on the other hand, manifests itself in innovative social behaviors – which, when they succeed, will be imitated, as the French sociologist Gabriel Tarde already explained a century ago –, and also in various typologies of economic enterprises, in exploration activities or in the generation
of new ideas. In fact invention is a kind of activity often based
on an abstract rational way of studying and controlling the
world by grasping and exploiting its uniformities through

means of mathematical models. Therefore, we should not forget that, as Poincarè rightly states,

'The genesis of the mathematical creation is (...) the activity in which the human mind seems to take very few from the external world, and in which it acts or seems to act only by itself and on itself (...)' (Poincarè, 1952).

In our approach to the artificial we shall concentrate on that particular field of activities which place at their center, on the basis of a more general imitation 'instinct', the reproduction of something existing in nature, and whose reproduction – through construction strategies which differ from the natural ones – man considers to be useful, appealing or in any case interesting.

Awaiting the new technologies which, according to Drexler's vision of nanotechnology, will allow us to

'build almost anything that the laws of nature allow to exist' (Drexler, 1986),

we shall take into consideration the efforts of men who try to reproduce natural instances through 'macrotechnology' strategies, on the basis of analytical models they build for such instances.

2   The Concept of Artificial: Fiction and Reality

From a linguistic standpoint, the term artificial (artificiale in Italian, kunstlich in German, artificiel in French) covers an ambiguous area which should be clarified before we proceed. In all languages, this concept seems to generically indicate all that is 'man made' and, at the same time, though more rarely, something which tries to imitate things existing in nature.

As the reader would naturally expect, we prefer the latter definition which, however, is not generally accepted. Nevertheless, it is a fact that, while no one would speak of an 'artificial telephone', everyone understands the meaning of an 'artificial flower' quite well. We believe that this situation can be interpreted quite easily. Though it has never been rationally defined, the concept of artificial refers to an object, process or machine which aims to reproduce some natural object or process. Since flowers exist in nature but not telephones, the adjective 'artificial' has no meaning if we attribute it to any object invented and built by man, i. e. an 'artifact', while it takes on full meaning when it is finalized to reproduce a natural object.

The Italian linguists Devoto and Oli have correctly defined the artificial as an object obtained by means of technical expedients or procedures which *imitate* or replace the appearance, the product or the natural phenomenon. Likewise, the imitation component is defined by the same authors as the capacity to get or to pursue, according to some criterion, varying degrees of similarity. The ambiguity of the question emerges, however, from the definition of the adjective 'feigned' which, according to Devoto and Oli, defines a product obtained *artificially*, as *imitation*.

Undoubtedly men, but also many animals, are familiar with the art of imitation and of deception (but, by the way, who would have accepted, for instance, the expression 'feigned intelligence' rather than 'artificial intelligence'?). Anyway, the semantic weight of this feature on the concept of artificial definitely seems to be too high.

The *perspectiva artificialis* of Leon Battista Alberti and Piero della Francesca – but also the landscape paintings of the so called Quadraturism, school born in the xvi century and enjoying success in subsequent periods (for instance, with Andrea Pozzo and his vault in Saint Ignazio in Rome) may be defined 'feigned', if you will, but only in the sense of something modelled, moulded by man as it is for the Latin origin of the verb *fingere*. However, the proper meaning of the term artificial is something that has often circulated in our culture in the most different fields, for instance in the following statement by Thomas Jefferson when he said

'For I agree with you that there is a natural aristocracy among men. The grounds of this are virtue and talents (...). There is also an artificial aristocracy, founded on wealth and birth, without either virtue or talents; for with these it would belong to the first class. The natural aristocracy I consider as the most precious gift of nature, for the instruction, the trusts, and government of society (...). May we not even say, that that form of government is the best, which provides the most effectually for a pure selection of these natural aristoi into the offices of government?' (Jefferson, 1813).

To sum up, though in every artificial object there is a deceptive or 'illusory' component by definition, it does not constitute its only component.

Thus, as described by Pliny the Elder, in the competition between Zeus and Parrasio, the former was so skillful in drawing bunches of grapes that the birds themselves were attracted to them; the latter, in turn, drew a sheet which seemed to cover a painting so realistically that Zeus himself was deceived by it.

Likewise, as described by Nicholas Negroponte, adding realism to an artificial system may sometimes have very strong effects on man too. When in the Seventies one of the first teleconferencing systems was designed in order to make the emergency procedures of the American government more efficient, a device was added to it, by means of which a moving plastic head indicated the person who was speaking at every moment, for instance the President. The result was that

'(...) video recordings generated this way gave a so realistic reproduction of the reality that an admiral told me that those talking heads gave him nightmares' (Negroponte, 1995).

In the above-mentioned examples and, overall, in the great intellectual achievements in painting during the Renaissance, it is very clear that the deceptive and 'illusory' component of the artificial, i. e. its 'fiction', is generated at different levels and with seemingly diversified meanings. Indeed, while in the case of painting the fiction is an intrinsic aspect of the object, in the case of the reproduction of the President's head cited by Negroponte, it is a secondary audio feature of the reproduction of his presence.

The famous desperate plea of that great sculptor who turned to his work and asked '(...) why do not you speak?', reinforces this point and the definition itself of artificial we are proposing here. Actually, the artificial, as an attempt to reproduce nature 'using different means', is seeking similarity, and, if it succeeds in achieving likelihood, deceives precisely because it is a matter of similarity and not of identity. Nevertheless, what is important is not the deception in itself but, indeed, the accuracy of the reproduction in the eyes of those who have to use or adopt it.

In this sense, as affirmed Prof. Willelm Kolff – one of the most important artificialists of this century who designed the first artificial kidney during World War Two and who then worked and still works in the field of the artificial heart – an artificial heart tends to 'cheat nature' because the blood it pumps arrives to the concerned organs 'as if' it had arrived from the natural heart. Nobody, anyway, would reduce such a device to a 'feigned' heart.

The fact is that the deception common language refers to, is usually associated with the artificial thanks to some of its external or 'aesthetical' appearances, like the aspect we can perceive in theatrical scenery, the crying of a traditional doll, the appearance of an architectural remaking by de Andrade or – but with greater caution because we are dealing with great art – the artificial perspective of a painting by Piero della Francesca.

In conclusion, the term artificial always implies the work of man. His 'art' in the broadest sense and the result cannot, therefore, but show traces of this origin: not nature but technology, even here, in the broadest sense of the word.

But this is only the *necessary* condition for discussing something artificial. In other words, an artificial object, process or machine is considered artificial because it is 'man made', but not all that is man made may properly be defined as artificial. In order to be truly artificial, an object has to satisfy a second condition, namely a *sufficient* one: it must be designed to reproduce an object existing in nature. Even the definition of artificial as something which is set against the natural is being called into question here. How would it be possible for the blood pumped by an artificial heart, to be used effectively by a natural organism if it came from an object set against nature?

On the contrary, the opposite of the natural is the conventional artifact, i. e. the product of conventional technology which, both in terms of material and, above all, in terms of its functions, leaves nature out. Actually, an object produced by conventional technology only exploits some laws of nature, is subject to natural constraints and adds itself to natural classes of objects or in any case acts on the natural world intentionally trying to change it. It is sometimes beneficial and sometimes creates problems of various kinds but those problems are different from those caused by nature.

The artificial, on the contrary, cannot exist without something natural that it refers to or tries to reproduce. The artificial, in other words, has a sort of umbilical cord which links it to nature and, therefore, it cannot ignore nature or, as a rule, attempt to change it.

3 'Copies' of Reality

Whoever has some familiarity with an electronic copier, knows exactly what is meant with the term 'copy': the reproduction of a document or an image on to another sheet of paper. The copy may be black and white or color, but, in any case, it is nothing but a photograph, at a given resolution, of the original document.

On the contrary, if one does not have such a machine at his disposal or a camera but only a sheet of paper and a pencil, then he could 'copy' the original text, summarize it or imitate the image by drawing a sketch just to render the idea. If we had neither a sheet of paper nor pencil, we could only try to memorize the relevant points of the text in question or the main features of the image.

In all three of the above-mentioned cases, the original remains what it was before, it was copied, summarized, etc. The only new reality will be the reality of the copy, our sketch or memory. Furthermore, the new reality resulting from our action will reflect the materials and procedures we have adopted. The resolution of the copier, and its ink, will modify the aspect of the original document or image in some measure, particularly if it is colored ink, while our style and our choices in summarizing a text or in sketching an image might even distort them.

In the third case, the 'materials and the procedures' we adopt, will be consistent with our imitation inclinations, our memorization abilities and related biases or habits we have acquired in our experience. Nevertheless, if the original text were a mathematical or chemical formula or an analytical report of an event – i. e. pure *information* – the accuracy of the copy, in terms of colors or drawing style, would be of little importance. As in the case of a train timetable, which could be written in very large characters on a wall or in very small ones in a brochure, the important thing is that the original information is rendered accurately and in its entirety. On the contrary, if our interest were just in the original as such, for instance in its graphic or esthetic style, then its information content would be less important. We could be tormented by the obsession, we see in collectors, to possess the original for its unique value and no copy would give us sufficient satisfaction.

In a word, man can only make copies of reality – which are identical to the original, i. e. duplicates or replications –if he is dealing with informational realities, such as computer programs, but not concrete objects. Mass produced techno-logical objects, where the prototype is reproduced by means of the same materials and procedures, constitute an exception which does not imply natural exemplars, but rather artifacts or systems intentionally and even formally designed by man himself. In the area of natural concrete objects – constituted by matter organized in a given way – replication is possible only through natural means and, of course, only where this is done by nature itself, as it is in the case of biological cloning controlled by DNA.

In every other area, man can only resort to the artificial. The artificial is not, therefore, a replication of reality, but a re-production, i. e. a production based on a natural *exemplar,* us-ing materials and procedures different from those used by na-ture. It must be noted that the constraints due to the materi-als and procedures are an unavoidable condition. Indeed, re-producing a natural object, e. g. a flower, using the same ma-terials and procedures, means replication, which, as we have seen, is utterly impossible at all. Frankenstein, therefore, be-longs to the realm of fantasy and not to the realm of bioengi-neering.

However, this does not mean, that man is unable to mod-ify nature acting on its own elements. As in the case of genetic manipulation, to cite just one example – but even in the more traditional procedures of crossbreeding animal or vegetable species – the recombination of fundamental structures of life is now within our capabilities.

Nevertheless, all this has nothing to do with the artificial because, in these activities, man acts within nature distributing, as it were, the same playing cards in ways not included in natural evolution.

In this sense, the understanding of the term artificial we are introducing here must be distinguished from the definition which has been widely accepted since Lucretius, according to which everything is 'artifice', because nature itself is able to make its own modifications, including those carried out by man. On the other hand, more recently, the chemist Roald Hoffmann, made a very persuasive argument that even so-called natural cotton does not differ so much from other synthetic fibers.

'A typical field of Egyptian cotton receives several treatments with insecticides, herbicides, and chemical fertilizers. The fiber is separated from the seed (ginned), carded, spun into a yarn. For modern shirting, cotton is also treated in a variety of chemical baths, bleached, dyed. It may be "mercerized", strengthened by treatment with lye (sodium hydroxide). Optical brighteners or flame reatardants might be added. Eventually the cotton is woven into cloth, cut, and sewn into a garment. It may be blended with another fiber for strength, comfort, or some other desirable property. That's an awful lot of manipulation by human beings and their tools, and to sharpen the point, manipulation by *chemicals*, synthetic and natural, going into your *natural* cotton shirt!' (Hoffmann, 1997).

These intellectual positions are quite convincing, but they are also largely reversible since, based on the same line of reasoning, one could maintain that everything is natural because everything is made of atoms and molecules. Such premises are of little use if one is aiming to gain an understanding of the possibilities, limits and consequences of human attempts to reproduce what he observes in the natural world; i. e., the results of variably complex combinations of atoms and molecules already generated, in given ways, by nature over millions of years.

The artificial, according to our definition, which is consistent with a part of the historical use of this term, consists in the result of human efforts to achieve the same results as nature using strategies which are different from those employed by nature, and, therefore, *lato sensu* technological. All this has nothing to do with, nor does it contradict, the thesis which maintains, *at a deeper level of analysis*, that all that happens in the universe is, by definition, internal and, therefore, natural (artificial included). Rather, even from this standpoint, it is possible to deduce that the artificial, though it is related to nature and without referring to nature it would make no sense, its intrinsic aim is to set up a new reality, a third technological reality inspired by nature.

In conclusion, it is reasonable to maintain that the artificial is always related to something which it is not and from which it draws its *raison d'être*. While nature is what it is, as it were, in absolute terms and conventional technology creates artifacts or processes compatible to a greater or lesser extent with nature but, in any case, not present in it, the technology of the artificial generates objects, processes or machines surely technological, but 'suggested' by nature. Therefore, at least ideally, the artificial should exhibit features which are not only compatible with nature but which clearly overlap nature.

On the other hand, a true technology of the artificial, as an autonomous body of knowledge, techniques and materials, clearly does not exist nor is it plausible to think it could exist. Actually, intrinsically artificial materials or techniques do not exist: rather, there are and there will be natural materials recombined by man using techniques derived from conventional technology and, let us say, forced to reproduce some natural object or process. In other words, the artificialist can only exploit materials and procedures made available to him by conventional technology. This is true for Icarus and the wax he used to fix his wings to his body and for the alimentary canal of the famous duck made by the 18th century mechanical engineer Jacques Vaucanson. Vaucanson was immediately interested when he heard about a gum from the Indies since it sounded to him like an ideal material to reproduce the internal tissues of his artificial animals (Fryer and Marshall, 1979; Bedini, 1964).

The same holds true for today's bioengineers who pay a great deal of attention to the discoveries made in the field of materials technology and in every other conventional technology, in order to find the most suitable component for their projects involving artificial organic structures or processes.

Artificialists have always adopted materials and techniques developed by conventional technology, even making some specific requests of technologists, to reproduce natural objects or processes.

Thus, in some measure, we are facing a sort of paradox. On the one hand, the artificial is intentionally related to nature since it aims to reproduce it. On the other hand, it inevitably depends on conventional technology, i. e. on a technology that, as we have seen, does not establish any reproduction target but, rather, builds objects, processes and machines which are heterogeneous compared to nature. As a consequence, the artificial swings between nature and conventional technology completely overlapping neither the former nor the latter. In the former case, it limits itself to natural replication and, in the latter, it distances itself inexorably from nature.

The destiny of the artificial, in whatever field it appears, cannot but bear the marks of this paradoxical ambiguity which, as we shall see, is the primary logical cause of its tendency to establish itself as a *sui generis* reality.

4   The First Step Toward the Artificial: Observation

The artificialist, i. e. the man– engineer, artist or whatever else he might be – who is attracted by the idea of reproducing something natural, is strongly characterized by a special way of viewing the world.

First of all, he must have a keen interest in observing nature, since it is from nature itself that he gets his ideas. Those who are dedicated to the design of artificial things should be able to grasp those aspects of reality which have a greater likelihood of being reproduced, just as scientists are sensitive to the perplexing and as yet unexplained aspects of what they observe and artists, in turn, concentrate on other aspects which allow them to interpret those aspects meaningfully.

Conventional technologists are also closely related to nature and, therefore, to its observation, but their main aim is to design objects, processes or machines which are able to control or modify natural events and not reproduce them. Conventional technologists often see nature as an adversary, while the artificialist looks at nature as a project to realize.

The four figures we have sketched, resort to four types or 'styles' of observation which are found, in varying degrees, in every human being, but are extremely relevant, above all, in the above-mentioned professions.

These types of observation are often interlaced – as in the case of Leonardo da Vinci – but they possess their own qualities and features. For example, a major advance in the development of the microscope was made thanks to a casual observation made in the xvii century, when, in order to enhance its magnifying power, the spherical-convex shape of a drop of water was used as a model. This discovery had already been made, without any subsequent development, in ancient times when it seems that spherical bottles full of water were used as magnifying devices. Of course, the observation of a physicist or chemist would be concentrated, of course, on the way a drop

of water forms, stabilizes morphologically, or on its internal
and external dynamics, while for those who were interested in
improving the microscope it was relevant as something natural
to be reproduced. Thus, innovators of the microscope are true
artificialists as was Watson-Watt, the inventor of Radar, who,
is said to have gotten the idea for his invention observing the
way bats detect obstacles during their flight.

These two examples allow us to highlight a basic truth.
Just as technologists of the artificial always resort to conventional technology in order to realize their projects, likewise
conventional technologists resort to the artificial in order to
develop or improve objects, processes or machines that, as
such, do not have reproduction aims.

In any case, it is clear that, to reproduce something, first
of all one must observe that thing and, as a consequence, observation is the first, unavoidable step every artificialist must
take.

But what exactly does observation mean? Discussing the
observation process would lead us astray down the often difficult paths of philosophy, but the problem in itself is not difficult to understand and the solution we propose is simple, at
least within this work.

As far as the observation process is concerned, the delicate point lies in its unavoidably 'relative' status. For instance,
everyone can observe the moon, but no one can observe it in
its entirety, or to cite a more relevant example of the relative
nature of the observation process, consider the observation of
a landscape. Clearly a geologist will observe a landscape in a
different way than an agronomist, a botanist or a painter.

We are then faced with another problem. Is what we observe reality or 'a part' of it, or an aspect, a profile which varies
according to the position we occupy in space or the dimension
– perhaps professional – we tend to privilege?

In this work, we shall define the profile we observe reality
from *observation level*. It is clear that the above-mentioned

problem is of particular significance not only for philosophers but also for scientists and, in a very particular way, for artificialists. When, for example, we design a human skeleton as an aid for teaching anatomy, which observation level should we adopt? Usually, a plastic skeleton will not encompass the molecular level nor the atomic level but will orient itself at the typical level of macroscopic anatomy. In other terms, if we dissected the bones, we would not be able to observe the biological structures that constitute bones in nature. On the other hand, even if an anatomy institute were very demanding and required that even the biological structures of bones be reproduced, using however materials which are different from actual bones, a threshold would soon be reached, beyond which one could not proceed, because of both our lack of knowledge and the difficulties involved in realistically rebuilding the connections between the various levels.

Perhaps, the painting by Jan Brueghel the Elder, *The Entry of the Animals into Noah's Ark*, 1613, explains what an observation level is better than a long discussion. The traditional representation of Noah's ark may help us to understand how man is naturally oriented to perceive reality in anthropomorphic terms: all the animals which were selected to survive have physical dimensions which are compatible with our senses, while microbiological living systems are damned to extinction. The love for certain animal species, furthermore, often imposes an observation level which is functional to the affectionate character that we attribute to that species. This polarization leads us to neglect those levels at which we could observe the brutality and cruelty of the observed species towards their prey.

In the field of artificial intelligence, no one would expect a computer program, able to reproduce the logical intelligence of a human being (e. g., the ability to perform correct deductions) that suddenly exhibit the capacity to compose poetry. In short, even intelligence can be observed and therefore described or modelized at different levels, and it is intrinsically very difficult to take more than one level into account simultaneously.

For the time being, we shall limit ourselves to asserting that human beings are forced, by their own nature, to consider only one observation level at a time. We can rapidly shift from one level to another, but each time we make a shift, we change radically the content of our observations, descriptions and judgments regarding what we observe. As we all know, a warm day can be cause for happiness or, on the contrary, a real problem according to the activity we had planned for that day (an excursion in the mountains or some strenuous job), though it is, in itself, the same climatic circumstance.

5 Eyes and Mind: Representations

The limitation of our observation capacities to only one level at a time, means that the artificial object which is made at the end of the design and building process, in no way will be the reproduction of the exemplar in its entirety. This would be true even if we could use the same materials and procedures used by nature. Though this case lies outside the field of the artificial properly said, even in the above circumstance we would be forced to use only those natural materials we have observed and not, of course, those remaining hidden and detectable only at other levels. This is a well known well-known fact for all those who have tried to reproduce some kind of fruit, flower or even some of their alimentary derivatives, using the same seeds and the same procedures nature does, but neglecting, or completely ignoring at all, other components – such as the composition of the air or the climatic dynamics – which make the development or the production of the exemplars possible.

As we have already noted, the need to use materials and procedures that are different from those used by natural exemplars, introduces a decisive constraint which prevents the artificial from approaching the exemplars beyond a certain point.

Nevertheless, we must again clarify another point concerning the observation process. Very often, we argue that, in the end, we see what we 'want to see', rather than some objective reality. This is an issue that it is well known to modern scientists – though in naïve terms but relevant for the history of science – in their anxious, and often frustrating, search for instruments and procedures for surveying reality 'as it is'. Just consider the following quotation of the founder of histology, the French scientist François Bichat, who, concerning the observation made possible by the microscope, said:

'(...) it seems to me that this instrument is not of great use to us, because when men look in the dark, everybody sees in his own way (...)' (Galloni, 1993).

Today, many philosophical doctrines focus on the process of observation in science, pushing positions that have regularly appeared in the history of thought for the past two thousands year to the extreme. In these doctrines reality do not exist at all, or, better yet, its existence has no relevance at the moment we observe it. What is truly important, according to these doctrines, is the act of 'constructing' the world, independently of its objective reality.

If we completely accept such a position then we should deny the objective validity not only of our daily observation but also of those made by scientists: everything would be uncertain, subjective and incomprehensible. On the contrary, it is clear that, at least in the field of natural phenomena, i. e. those phenomena that can be precisely measured using instruments, the world has its own objective capacity to act on man, starting with its actions on our senses – just consider an earthquake. In this case, the epistemological position of the observer in no longer relevant.

If we referred to the constructivistic doctrines, it is only because they, at least, remind us of the fact that nature and its events, as we have already underlined, are not at our disposal in an 'immediate' way. Our knowledge of the world is 'mediated' by our mind, the place where we form *representations* of reality. A representation is the mental reproduction of what we observe through our senses or something we generate autonomously, for instance the image of a face or lake, but also our subjective description of the atom or of a continent, and even our 'vision' of the universe or God. Therefore, a representation is something 'meta-artificial', a sort of prelude, as it were, to the artificial: it is, in fact, a non-material construct which reproduces the world we observe and is definitely useful for surviving in it.

Without representations we would be not able to evaluate situations or associate memories and observations: we could only rely, each time, on immediate reactions to reality as many animals surely do. Fire would burn us every time, because, within our mind, we would have no recorded notion of its features and effects. In practice, when we encountered fire, we would not recognize it, and, likewise, its symbol drawn on a wall – as a result of a collective representation – would have no meaning for us and we would not be kept on alert. In short, the representational system adopted by the subject gives meaning to his action and this meaning may evolve during the problem solving process (cfr. Leiser, Cellèrier and Ducret, 1976).

Forming representations is closely related to our perception and observation and different classes of representations correspond to each level,

The mind does not play a passive role in forming representations because, just like in the choice of an observation level, the whole system of our experiences and our preferences, interests and fears, acts on it. Culture itself is a powerful source of 'directions' we should follow, or refuse to follow, in observing reality and, therefore, in forming representations. For instance, the sociological or the psychological dimensions of man are today widely accepted as a reality, but it is a matter of a relatively recent fact.

Of course, man has always had, what today we call a sociological or psychological dimension of life, but the lack of representations, models and theories, specifically dedicated to these phenomena, did not appear until the xix century. This scientific void prevented those dimensions from becoming observation levels, together with daily, political or spiritual observation levels.

The same holds true for other observation levels: the subatomic or the ecological, the economic or the magnetic, the micro-biological or the chemical, etc. Each of these levels is, simultaneously, the cause and effect of representations which, almost referring to the same natural object or process, privilege one profile only, only one way to be or to present itself to the observer.

Drawing further upon Heisenberg, we can speak of a sort of generalized indetermination principle: the choice of an observation level allows us to capture a true reality, but only the one which is compatible with that level. In more general terms, one could say that in selecting an observation level,

'(...) we force the matter to choose a configuration from among those that are available (...)' (Regge, 1994).

These considerations are truer than ever, for all the arts. Perhaps is why Oscar Wilde maintained that art is nothing but a form of emphasis, exaggeration of the reality we perceive. Though, since Aristotle, art has been assigned the role of imitating nature, *de facto* no artist can generate nor does he wish to generate copies of what he observes, but, rather, an interpretation of it according to his own poetics, i. e. his own representational modalities, or those of the school he belongs to.

The history of painting, for instance, exhibits a very wide range of pictorial observation levels which express representations that are very different from each other, even when they deal with the same 'subject': from the spirituality of the Middle age to the concreteness of the Renaissance, from the vagueness of Impressionism to the complexity, often solipsistic, of the vanguards (Bertasio, 1996).

Indeed, it is through art that we can understand to what extent mind and culture may sometimes be active protagonists in forming or in confirming many orders of representations. Just consider painting in the Middle Ages, the pictorial reproductions of God or the devil – for instance, the chilling reproduction of the devil by Coppo di Marcovaldo in Florence. This demonstrates that, at one extreme, man is able to generate artificial objects based on exemplars which are quite inexistent in nature, but, notwithstanding, accurately represented in cultural traditions. Likewise, the current representation of the atom, with its analogy with the solar system, has become a true graphic symbol widely shared by scientists and common people, though it is very far from being verified in its structure, dimensions and dynamics. However, though it is extremely useful and necessary and makes the research in the field of physics possible.

6 The Exemplar: Background and Foreground

As we said, the exemplar is the object or process which is chosen as the target of the reproduction. More precisely, we should say that the artificial is the reproduction of the representation of the exemplar, which the artificialist generated in his own mind. Models, even purely mental ones – in the technological design as well as in art – are examples of representations which function as pilot images, maps or schemes of the natural object or process which we wish to reproduce.

Even the choice of the exemplar, however, is not a simple task, devoid of ambiguities as it might first appear. We all know that, in considering an artificial heart, we are referring to a well-known and recognizable exemplar, that would appear to be easily distinguishable from all that is not a heart. Obviously, for an engineer the question is much more complex: which organic parts, vessels, muscles, sub-systems, define the heart? In other words, what are the 'boundaries' of a heart?

In addition the realization of heart valves, today there are also devices which aim to re-create the left ventricle (the so-called *left ventricular assist systems*) and are designed to work together with the natural heart of the patient. Other devices reproduce both ventricles. The total artificial heart, able to completely replace the natural heart, has only recently become an achievable aim, but many problems remain. Often these problems are related to the sub-system – which, therefore, must be considered as a part of the exemplar – dedicated generating electrical power for the control of the electronic circuits and moving parts.

In another field, suppose we wished to reproduce a pond. How should we establish its boundaries? On a topological level, should we even include the geological structure the pond's bottom and sides or not? As far as the flora and fauna are concerned, how close should we come to the natural pond? Should we include every aspect of the pond's ecosystem, which ranges from ducks and fish to microbiological creatures? It is quite clear that, different answers to such questions will give rise to different models and concrete reproductions of the pond, and these differences may be critical if related to our aims.

The same is valid for artificial lakes, because they

'(...) are typically much shallower than natural lakes – explained Charles Goldman, professor of limnology at UCD in the department of environmental science and policy, and director of the Lake Tahoe Research Group – (...) They're often so shallow that they do not stratify, or circulate from top to bottom. (...) Which means they're excessively productive – green, essentially. Artificial lakes are particularly hard to manage (...) like an algal generator. This lack of stratification, along with the fact that typically, such lakes are surrounded by fertilized lawns or function as drains for storm waters – causing excess nutrients to flow into them – can result in eutrophication' (Goldman, 1999).

In the field of artificial intelligence this is a well-known much debated problem: how may we define – a verb which, by the way, derives from the Latin *definire*, which means 'to fix the boundaries' – human intelligence in comparison with other functions of the mind like memory or intuition, fantasy or curiosity?

At one extreme, we could consider exemplars from the animal field, such as the *holothuria* (it is also known as the sea cucumber) which lives symbiotically with the little fish *Fierasfer acus*: how could we separate these two entities, first of all in representational terms and then in terms of design and reproduction?

It quite clear that the task of outlining an exemplar consists in an operation which is always arbitrary to some extent: a true isolation of an object or process from a wider context which includes it or from an environment which hosts it.

Western civilization, because of its philosophical and scientific tradition, has demonstrated an ability to carry out 'analyses' of the natural world, greatly benefiting from such analyses. But analysis – a term which, interestingly, derives from the ancient Greek 'to break down' – is surely much more useful for scientific than for artificialistic aims. In fact, while the knowledge we acquire through analysis is always to be considered as valid, at least in descriptive terms and, sometimes, even in predictive terms, the reproduction of an exemplar which, in nature, behaves in a given way, could require the cooperation of many parts of that exemplar. This will require, in turn, more than one observation level, and the analysis, with its usual isolation strategies, might not be able to make all the required levels observable.

In other words, the choice of an exemplar is a sort of literal 'eradication' of some part of nature and this can take place, as we saw, both in terms of a true isolation in space, and in structural terms. On the other hand, man seems to have no alternatives: as we see in the simplest daily observations, or even in observations aimed at defining an exemplar to reproduce, man cannot but proceed by putting one thing at a time in the foreground and relegating to the residual things to the background.

All this happens as a function of the choices imposed by observation levels. Hence, the functioning of an artificial device will be similar to the exemplar only to the extent that the observer, or whoever has to use it, evaluates it by placing themselves at a level which is as close as possible to the level of the designer.

Of course, if the observer and the user have to deal with an artificial device conceived and built based on an observation level that has never been experienced or that is extremely subjective, then they will have to face many additional problems in order to evaluate the quality of the reproduction.

This is true not only for artistic reproduction – which is innovative by definition – but also for scientific innovation process. Actually, the reproduction of the solar system according to a Copernican representation, by means of a mechanical device, would not have been easily understood and appreciated at a time when the Aristotelian representation of the universe was the commonly shared model.

On the contrary, under the influence of the mechanistic culture of the xvi and xvii centuries, Kepler was able to develop a three-dimensional reproduction of the universe, the famous *Machina mundi artificialis*, which was subsequently given to Prince Friederich Von Württemberg.

Thus, defining the exemplar is a task which, though it appears to be quite obvious in daily life when we indicate the objects or the events that surround us – or happen within us – proves to be quite complex as soon as we try to begin a project or design intended to reproduce it.

One of the lessons to be learned from the study of the artificial is our own limitations, first of all, in knowing nature, which is a phase that precedes – or should precede –the phase of reproduction.

Analysis often allows us to control reality, such as when, for instance, we succeed in describing with some accuracy the anatomy, physiology and perhaps even the psychology of a given animal, e. g. a pigeon. Thus, we can explain many of its behaviors, pathologies or abilities. This knowledge, however, is not sufficient for designing an artificial pigeon, not only because it is only a fraction of what we should know about a pigeon, but also because it has been acquired, unavoidably, through analytical strategies. These strategies, break reality into areas which are often heterogeneous as the observation levels they derive from, and their synthesis would require knowledge that could only be acquired through as yet unknown strategies.

7   Essentially, What Is a Rose?

The choice of an observation level and an exemplar are the first two steps in the process which ultimately leads to the design of an artificial object.

Nevertheless, the choice of an exemplar is not the final conclusive defining moment of what will be done. As soon as the problem of the delimitation of the exemplar – which we discussed in the previous section – is solved, we are faced with a new and decisive step. This moment could be defined as the choice, or the attribution, of an *essential performance* for the exemplar.

Essential performance is the quality, function or behavior of an exemplar or even simply the aspect of the exemplar that the artificialist believes to be particular or typical and which cannot be omitted.

In the previously mentioned case of the heart, there is a general consensus on what its essential performance is, i. e. the pumping of blood. This performance of the natural heart is fundamental: a design of an artificial heart cannot neglect such a feature. Indeed, an artificial heart that, at the observation level we assumed, namely the physiological one, was not able to pump blood would be quite unrecognizable as a heart and, more importantly, useless.

From an bioengineering standpoint, there is general agreement on what must be considered as essential in the case of the heart. There is consensus regarding kidneys, lungs, pancreas, liver, skin, some tissues, etc. However, this kind of accord is not found in all areas of the artificial.

In the field of artificial intelligence, for instance, since re-searchers have to deal with a rather unknown exemplar – the human mind – there are a number of different opinions regard-ing its essential performances, and these viewpoints are often incompatible with one another. Intelligence itself, which, at first glance, might seem the essential performance of our mind, is far from being accepted by all researchers as the main mental characteristic, at least if, by intelligence, we mean the ability to solve problems. Besides other types of non-formal in-telligence – such as concrete or motor intelligence – we should take into account creativity, the ability to recognize objects or situations, memory, attention and many other functions.

Thus, it can be asserted that, in the field of artificial in-telligence, there are as many schools of research or design as there are possible viewpoints concerning the essential per-formance of the mind, though all deal with the same exemplar.

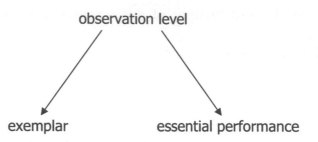

The multiple selection model involved in the design of an artificial object.

The main schools, which have difficulty finding common ground, are the so-called schools of symbolic artificial intelligence and neural networks. The former insists on the design of computer programs that try to reproduce intelligent behavior on the basis of algorithms. These algorithms which simulate mental representations and knowledge through quality and numerical symbols and through logical evaluations of these symbols (deduction, comparison, association, calculus, etc.). The latter, on the contrary, draws the long-neglected cybernetic works of the 1950s, and aims to design devices whose intelligence consists in the automatic recognition – neither logical nor symbolic –, on the basis of suitable 'training phases' of the net and series of diverse data: meteorological situations, geometrical shapes, several kinds of objects to be detected in ambiguous and confused contexts.

Furthermore, the attribution of an essential performance is always a process in which empirical reality and the autonomy of the mind overlap. Attribution of a certain performance to a ductless gland may be rooted in a way of perceiving life proposed by some established theory, the premises of a religion, or other subjective preferences of the researcher.

In every culture, the conception of man himself, is of course also based upon a multiple inclusion of essential performances. Philosophical research, regardless of the particular school of thought, is proof of the enduring effort to discover the essence of man.

Especially evident is the inclusion of selections, or attributions, in the following Hindu axiomatics contained in the Chandogya Upanisad, in which one can find the

'(...) various steps that mark the subsequent materialization of the world: the saman is the **essence** of the poetical meter, the meter is the **essence** of the language, the language is the **essence** of the man, the man is the **essence** of the trees, the trees are the **essence** of the water and the water is the **essence** of the earth' (Schneider, 1960, emphasis added).

Once again, in the field of biology, botany has shown that, throughout history, a strong inclination to generate attributions of changing essential performances is often dominated by extra-rational visions. We can see an example of this phenomenon in *Malva silvestris*, which, according to its classification by Limneus, has been used since antiquity for its medical properties, and which, according to the Pitagoreans, also boasts the capacity to save human beings from being slaves to their passions, while later, in the eighteenth century, it was appreciated for its alimentary virtues; another example is *Cheiranthus cheiri*, which, again according to its classification by Limneus, was used by the Greeks and the Arabs as a cleaning substance and its cardiotonic properties were only later discovered in the twentieth century.

The choice of an essential performance is a moment in which, the artificialist may be as aware as he ever will be of the fact that, by privileging one performance over others – which are in principle unlimited in number – he always acts arbitrarily, and is strongly influenced, as we have already illustrated, by the observation level he assumed at the beginning of his work.

Even the above-mentioned Vaucanson was aware of this when, speaking of the digestion of his artificial duck, he defined the essential performance he wanted to privilege as follows:

'I do not claim that this should be perfect digestion, able to generate bloody and nutritional particles in order to allow the survival of the animal. I only claim to imitate the mechanics of this action in three points: in the swallowing of the wheat; in soaking, cooking or dissolving it; in allowing its going out forcing it to visibly change its stuff' (Vaucanson, 1738, seen in Losano, cit., p.91).

Two American researchers, M.A. Mahowald e C. Mead, among the many who are engaged in the design of an artificial retina of the human eye, made similar observations:

'In building a silicon retina, our purpose was not to reproduce the human retina to the last detail, but to get a simplified version of it which contains the minimum necessary structure required to accomplish the biological function' (Mahowald, C. Mead, 1991).

Nevertheless,

'The real vision (...) will probably require that artificial retinas contain 100 times the number of pixels and auxiliary circuits, to imitate the functions of perception of the movement, and to intensify the contours performed by the amacrine cells and by the ganglion cells. Finally, these systems will also include additional electronic circuits for recognizing configurations generated by the retina' (ibidem).

Even in a therapeutic approach, in the field of optometry, there is a great need for artificial devices. Just consider treatments for dry eye syndrome, and

'(...) there are formulation called artificial tears that are available in the market today. They are only partly satisfactory. (...) we need to dig in and ask what the various components of the tear fluid are, and which of the components are indispensable and we cannot do without, which are those that we can do without, but it would be nice if they were there but there is no harm if they are not there in great amounts, and whether there are components in the tears that we can do away with. The reason this question is hard is because it

has been estimated that there may be a hundred different components in the human tear fluid. When we want to formulate an artificial tear fluid, we would want to do it with a number far smaller, and
yet the fluid that we create should be able to satisfactorily do the job'
(Balasubramanian, 1998).

The same holds for the cornea and vitreous.

Another typical case, which is analogous to the previous
one, is the attempt to reproduce the propulsion used by fish
with a robotic model. Until now,

'It is almost impossible to reproduce the performances of fish
simply by imitating their form and function, because a vehicle able to
set up uniform and continuous flexes, having a body similar to that
of a fish, is quite beyond the state of art of robotics' (Triantafyllou E
Triantafyllou, 1995).

Nevertheless, it is possible to imagine that

'In the future, such creations which are inspired by nature, will
perhaps improve their biological models for some specific task, like,
for instance, the exploration of the sea-bottom.' (ibidem).

In its general terms, the question posed by the selection
of an essential performance can be outlined in the following
way, drawing from the possible design of an artificial rose. In
this flower, as in every other biological system, – neglecting the
problems of its delimitation in the space and its structure,
which we shall consider as decided – what is essential?

It is quite clear that our answer will strongly depend on
the observation level we assume. The physical observation
level that we have established (chosen or constructed) in order
to indicate the flower as an exemplar (e. g. micro, macro or
some intermediate position) will lead us to attribute some essential performance to the flower (e. g. shape, color, fragrance,
some kind of behavior, etc.) and to neglect other possibilities
(e. g. the consistency of the tissues, the cells structure, the
system of wessel, etc.)

A commercial manufacturer of plastic flowers will probably select an essential performance in terms of pure appearance, while a manufacturer of educational tools will concentrate on the structure, usually only macroscopic, of the main anatomical parts of the rose. A publisher, in turn, for illustrative aims, will have an expert create an image that outlines the some drawing able to outline the same parts of the rose on the printed page, therefore in only two dimensions, while, at the opposite extreme, a painter will decide according to a much free range of interpretations.

It is interesting to note that in children's drawings the essential performances are never as simple and preponderant, though often one needs an explanation by the child – as is the case for abtsract artists – in order to understand the graphic language and, therefore, the theme at the drawing. The fact is that, during the age of socialization, the ability to simplify reality first begins to develop. It will later give rise to the selective attitude we are speaking about, useful not only in making the artificial but also in operating in the world of science and technology and, in the end, in surviving in everyday reality itself.

8   Reality Does Not Offer Any Discount

In 1983 in one of the rare works, not Italian, dedicated to the artificial, the biologist Martino Rizzotti though orienting his discussion towards an understanding of this concept, which is rather traditional (as anything made by man), insightfully notes that:

'(...) whatever intervention (action, technological construction, etc., NdR) involves, for its own nature, an amount of mass and energy which concerns not only the resulting object but also the environment. (...) If we displace some stones we displace always some amount of mould, we crush some insect, and we lower the soil (...).

Even if these effects are microscopic and secondary, they appear always along with our action' (Rizzotti, 1984, 34).

This emphasis allows us to introduce, after having become familiar with the three main concepts of the theory of the artificial – the observation level, the exemplar and the essential performance – a principle which is fundamental to our discussion: the *principle of inheritance.*

This expression refers to a circumstance that is very simple in itself but often neglected in theory and practice. Whatever action we perform, including the development of an artificial object or process, is not limited to the effects which are predicted and planned by the design but includes many other effects regarding of a quality and quantity which are not predictable *a priori.*

If you read the description of any drug, you will immediately see what we are talking about. *Side effects* are a very important part of the work of a pharmacologist, who, having to pursue objective O_1, knows perfectly well that his drug will also generate objectives O_2, O_3, ... O_n, (where 'n' is an indefinable quantity) which are not part of his original objective.

Side effects frequently manifest themselves in the form of unpredictable *sudden events*, caused by particular combinations of events which are sometimes explainable but often unexplainable. Such events are common not only in the area of pharmacy and medicine but also in meteorology, engineering and many other fields, not to mention all natural or technological dynamics.

The main point, as far as side effects related to technology are concerned, is that technologists – but this is also true for every man – cannot do actions which, having to achieve some objective, are able to achieve that objective and only that objective.

In other words, whatever action we do, does not just generate the effects it was intended to generate. For instance, when someone buys a car for the first time, the car will trigger a series of new events which go beyond his planned aims. As we know, the use of a car changes our habits, provides us with new sensations, can affect our physical appearance, implies a revision of our budget, absorbs time for its maintenance or repair, contributes to pollution problems, etc. It is not simply possible to buy a car and nothing else. Its 'side effects', those which are already known and those which will come as a surprise, cannot be eliminated.

This is also why, for instance, music may have no meaning in a strictly semantic sense (Sloboda, 1985), but it surely has extra-musical effects. For example, just as a drug may produce biological dynamics which are different from the action level at which it was designed and tested, music may have an effect on some classes of physiological phenomena.

We could say that reality, as it were, does not give a discount and that our actions always have numerous consequences, many of which are impossible to predict. These consequences will involve several observation levels which are generally not considered when we plan our actions.

In the case of the artificial, the inheritance principle is made possible by the subsequent choices that the artificialist will inexorably make (selecting an observation level, isolating an exemplar, privileging some essential performance) and also, perhaps above all, by the materials and procedures he decides to adopt.

The role of multiple selections and the inheritance principle in giving the artificial its own properties, is well illustrated by the following anecdote, told by the art psychologist Rudolf Arnheim:

'The smoke detector in the new library where Mary works proved to be so sensitive in the beginning, that on two separate occasions, when an employee lit a cigarette in the office, firemen were called. Some sensory devices artificially created by man, respond to danger signs with greater reliability than the the senses we are born with' (Arnheim, 1971).

It is clear that no human would have acted in such an exaggerated way, but an artificial device, lacking any discriminatory capacity and privileging the essential performance for which it was designed and only that function, is inevitably vulnerable to such reactions. The same is true of so-called 'errors', which often characterize computers, burglar alarms and many other control devices.

Nevertheless, the artificial exhibits its greatest exposition to the inheritance principle when the materials are concerned.

This issue may be summed up in very simple terms. On the one hand, we have no *a priori* opportunity to realize how many observation levels constitute natural reality or, in any case, which ones we should adopt in observing it. Such a list would end up being arbitrary, but, above all, definitely incomplete. For example, how many levels define a stone?

The list could begin at a strictly petrographic level and then move on to the physical macroscopic and chemical levels, including the geological level, proceeding then to the physical microscopic, atomic, electronic, etc. Thus, when we use a stone in a given technological project (in the field of the artificial or in a conventional technological field) we do so because some of its properties appear suitable for some aspect of the design itself.

This property, however, do not exhaust the 'nature' of the stone in question. The property which would attract the attention of the technologist, in fact, will be perceivable at some observation level, but, if we shifted to another level, then we would discover dozens of different properties of the same stone. Some of these hidden properties could, *a posteriori*, make the achievement of the planned aim of the design easier; others, on the contrary, could prove to be obstacles while others might be neutral.

Inevitably, when we adopt the adoption of the chosen material, of that stone, we will *inherit* all its properties and not only the one which attracted our attention. The inherited properties, may remain dormant silent for an unpredictable amount of time, revealing their presence and affirming their 'rights', in special circumstances which are quite unpredictable. The impossibility of their *a priori* description depends, as it is easy to argue, on the number of possible interactions which the various levels of the stone could maintain with the levels of the real environment the artifact is hosted in.

In the end, the number of the possible interactions cannot be calculated because, if we assume that any natural object is characterized by infinite observation levels, then its encounter with whatever real object – which is in turn characterized by infinite potential levels – will produce a quantity of possible interactions which is equal to the product of two infinite numbers, which is an undetermined number.

Remaining in the field of construction materials, it is well known that some marbles are more sensitive to rain and to its chemical components than others, which are more sensitive to other organic or inorganic natural phenomena. In all cases, the result is that, after a period, the appearance and sometimes the structure itself of the building is strongly modified by these undesired reactions. Mechanical or chemical procedures adopted for cleaning such buildings, in turn, lead to new problems. Thus, as has been observed in some studies carried out by the Masonry Conservation Research Group of the Scottish Robert Gordon University, in some circumstances buildings have been damaged to such an extent that the stones rapidly decayed. Indeed, according to the study, cleaning work was carried out quite ignoring its effects and the consequences.

Even artificial marble, which dates back to the Baroque period in Europe, needs to be cured for, because of

'(...) degradation and weathering of artificial marble by environmental and climatic elements. Water and salts from rain and rising humidity from the ground, by deposition of aerosol and gases, as well as changes in humidity and temperature are responsible for weathering of building stones and plasters. The main reasons are hygric and thermal swelling and shrinking processes as well as crystallization of salts' (Wittenburg, 2000).

Current plastic marble imitations certainly would have their own story to tell in this regard.

Let us consider some other examples. A contemporary operating room, unlike those of past centuries, is a very controlled environment because today we know that whatever object or surgical instrument is used by the surgeon, it does not just possess the desired properties (for instance, mechanical). It may also have other are rather dangerous properties, including microbiological entities, dust, metallic residuals, etc. However, despite all the precautions which are taken, no operating room in the world may be said to be completely controlled, since, by definition, we can only control the phenomena we know and, sometimes, only partially.

The most evident phenomenon, in a medical or biological field are the various kinds of so-called 'rejection', including reactions of the immune system. Rejection is a true rebellion of the organism not against the essential performance that, for instance, a layer of artificial skin may exhibit for a wound, but against some component of it which is perceived as extraneous or dangerous to the body.

Dr. T. Keaveny, from Berkeley, sheds light on the issue in describing the problems that bioengineers face in their attempts to build and to place artificial bones into the human organism:

'(...) joints are trouble-free for 15 years (may be evaluated as) a remarkable record considering the harsh biomechanical and biochemical environments of the body' (Keaveny, 1996).

Because of the immune system reactions other researchers of Rice Institute remember that until recently:

'(...) most research in the field (of cell transplantation, n.d.r) has focused on minimizing biological fluid and tissue interactions with biomaterials in an effort to prevent fibrous encapsulation from foreign-body reaction or clotting in blood that has contact with artificial devices. In short, most biomaterials research has focused on making the material invisible to the body' (Mikos et al., 1996).

A biomaterial has been defined by the American National Institute of Health, in 1982, as

'(...) any substance (other than a drug) or combination of substances, synthetic or natural in origin, which can be used for any period of time, as a whole or as a part of a system which treats, augments, or replaces any tissue, organ, or function of the body' (NIH, 1982, 1-19).

The same Institute underlines that

'Materials science was defined as the science which relates structure to function of materials.(...) The field of biomaterials is first and foremost a materials science' (ibidem)

and that

'In evaluating safety and effectiveness of biomaterials, the material cannot be divorced from the device. (...) Each biomaterial considered for potential clinical application has unique chemical, physical, and mechanical properties. In addition, the surface and bulk properties may differ, yielding variations in host response and material response' (ibidem).

It is important to note that, though both the statements by Keaveny and Mikos were made in the second half of the 1990s, bioengineers are currently experimenting with new strategies for getting biomaterials which may be called 'hybrid'. This means that the new materials aim to harmonize the artificial organ and the host environment.

Using suitable biomaterials, they try to give, for instance an artificial cell, sufficient compatibility with the host organism at the surface level, while maintaining the needed artificiality in its internal structure. This is another example of an attempt to deceive nature, since the organism will be induced to accept the performances generated by the artificial cell without attacking it, because its interactions with the cell will be mediated by a compatible surface. 'Hiding' is another expedient.

'(...) insulin-secreting cells form pancreatic islets, usually taken from a pig, in semi-porous capsules implanted in the body. The capsules must be biologically and chemically inert; that is, their chemical composition cannot induce inflammation or other reaction from the body, and they must resist decomposition. The capsules must contain pores small enough to exclude the mobile cells of the immune system, macrophages and lymphocytes, but large enough to allow a physiological release of insulin in response to blood glucose levels' (Edwards, 2001).

The possible extension of such a hybrid strategy to other areas of the artificial raises interesting possibilities, though not completely new ones, if one considers, among the many cases we could cite, the attempts to make artificial intelligence programs or robot something friendly or even anthropomorphic (that is to say, externally similar to human beings by some kind of interface). In all cases in which designers resort to such interfacing strategies – between the artificial device and the natural world –, their meaning should in fact be found in the effort to make compatible two heterogeneous realities, at a macroscopic perception observation level, whose interactions are controllable only within a very narrow range.

Regarding the functioning of artificial cells, McGill University researchers (where Thomas Chang began this kind of research in 1957), claim that artificial cell membranes can be significantly modified by adopting biological or synthetic materials. Their permeability can be controlled in many ways. In this way, the materials enclosed in the cell can be held back and kept separate from undesired external materials. On the other hand, Chang himself said in 1996, in contrast with de Monantheuil's statement cited at the beginning of this book,

'(...) if you look at it objectively, no matter how smart we are, we will never be able to copy what has been made by God, not even a simple red blood cell. We can only hope to make a simple substitute, and right now we are still taking our first steps' (Chang, 1996).

We can also interpret along the same lines the attempts, which appear very frequently throughout the history of technology, to build automata – just consider to the copyists of the Swiss Pierre Jaquet-Droz and many other examples of the xviii century (Bedini, 1964). These attempts, furthermore, converge with all the tradition which go back to the mannequins universally used in many kinds of theatrical performance and the like, or rites, or, today, for emulating situations and events in medicine or in the military.

Though the issue is somewhat extraneous to our discussion, it would be very interesting to ask ourselves what relationship might exist between all this and the very ancient inclination of man to build camouflage devices to alter his own identity, aspect or presence, dating back the most ancient masks used in the funerary Egyptian art or in the Greek theatre.

In any case, it is clear that every concrete attempt to reproduce a natural exemplar, and a performance of it which is considered as essential, implies the generation of a set of realities much richer than one might desire.

9   The Difficult Synthesis of the Observation Levels

The situations within which scientists and technologists act, above all those regarding the technology of the artificial, are surely complicated, permanently, by our incapacity to place ourselves simultaneously at more than one observation level.

Though scientific methodology has developed several techniques for controlling more than one variable, *de facto* our theories and models always unavoidably focus on some aspect or profile that is always considered as central. All you have to do is consider the history of recent science to realize how true this is.

The fact is that, attributing some privileged and comprehensive meaning to the events we observe at a given level is a limiting process which prevents us each time from looking these events from other levels if not in different moments.

It is a 'temptation' that has appeared many times in the history of scientific ideas: for instance, in the past century, were attributed central roles in a number of phenomenologies, the sexual act included, to electricity (Leschiutta, Rolando Leschiutta, 1993).

Today, the dominant theme, which we have already referred to, appears to be information, to which is reduced a series of phenomena such as intelligence and communication, the intimate stuff of social relations and even art.

In short, the XVII century was largely dominated by a mechanical observation level, whereas the XVII and XIX were dominated by electrical and information levels, respectively.

The history of the damage or, at the very least, the waste in terms of research projects, caused by these reductions or polarizations, has yet to be written, but, in the end, the human inclination that produced them – that of privileging one observation level at a time – definitely seems to be without alternatives. If we consider the world of the artificial, the above-mentioned inclination goes hand in hand, as we have seen, with another unavoidable constraint: the one which forces the artificialist to use different materials and procedures than the natural exemplar and, thus, introducing to the object or process which is to be built, a tendency to generate side effects and sudden events whose frequency, intensity and quality are unpredictable.

In the best case scenario, it is only possible to limit these effects, as we have seen in the previous section, by means of a sort of 'encapsulation', i. e. isolation of the artificial from the external world, for instance the host organism. In such a way, the only interaction which can take place – between the artificial and external world – is the one we defined as essential performance, the only umbilical chord connecting, to some extent, the artificial and the natural.

Now we can pose another question. In general terms, under what conditions and with which results, could one design more exemplars and more essential performances which cooperate in a unique system that reproduces some natural system?

If we take the case of a flower which we want to reproduce at more than one observation level by means of materials and procedures different from the natural ones – for whatever aim: scientific, educational, commercial, etc. – then we would face a fundamental difficulty. For instance, we must decide what kind of relationship among its parts (stamen and carpels, style, ovary, stimma) we should reproduce. In other words, at which observation level (cellular, molecular, etc.) should we reproduce the relationships which characterize the flower as whole, in order to accurately reproduce not only its structure but also its functions?

Clearly, presented in this way, the problem is theoretically and practically irresolvable, since, if an artificial flower must be a reliable anatomical and physiological reproduction of the exemplar, then, if we succeeded in this sort of reproduction, we could claim that we have replicated it, i. e. recreated it artificially entirely. The analytical rebuilding, piece by piece, of a living system – but also any other sufficiently complex natural reality – starting from its basic chemical elements, is a task which is definitely beyond our capacity and, perhaps, intrinsically impossible. Indeed, genetic engineering may be able to achieve this kind of reproduction, but, of course, in this case, we do not rely on different materials or analytical models of the various sub-systems the whole system should contain. In genetic engineering we only set up the conditions through which nature works by itself.

But artificialists have a very different aim. In principle, what they are attempting is, the replacement of natural materials with other materials which approximate them at a given observation level. Our current knowledge of organic and inorganic materials and our ability to manipulate their features – enhanced a great deal by space technologies – allows us to generate substances and physical or chemical structures, in many fields, which are very similar to the natural ones. However, this similarity is almost always recognizable at the selected observation level.

Returning to our case of the flower, we are able to generate a wide range of artificial scents and, among these, we could select the one which is the essential performance of our exemplar. But, if our aim is to rebuild even only the structural and physiological sub-system which produces that particular scent in the flower, then we are faced with quite a different problem. We must decide what relationship to establish between the artificial scent and the artificial structures we would have set up, in order to allow those structures to generate our artificial perfume, rather than simply pouring it on them.

At this point, it is clear that, the design of the partial artificial objects (scent on the one hand and some anatomical structure on the other) would be contorted, since their reciprocal dependence in the new possible design will surely impose an indefinite quantity of adjustments.

A huge set of new problems would the arise, and, these problems, would require some drastic decisions. For example, we might decide to establish an anatomical threshold under which we might give up the criterion of similarity with the exemplar, thus limiting the reproduction to some more superficial aspect suitably described by a reasonably simplified model.

As usual, in the artificial flower as well, some artificial device would be made to generate the scent while remaining invisible within the whole organism. Even if its characteristics would were compatible with the system, other artificial parts might be forced, through some expedient, to come in contact with the scent in an a way sufficiently similar to the natural one, at least, in terms of the new observation level which we shall have generated or selected.

It is important to note that the observation levels at which the two partial artificial objects are reproduced, will no longer have any relevance in the sub-system we have built up. They will be 'absorbed' by a third level: the level at which the relationship between scent and anatomic-physiological structures of the flower become possible. The two original exemplars and related observation levels will be, as it were, sacrificed for this relationship. They will be re-modeled in order to serve the new essential performance, which is constituted exactly by the relationship between the two exemplars.

As we have seen in the unavoidable complications of the above-mentioned logical procedure, the cooperation between two artificial objects or processes poses serious methodological problems. Man, almost invariably, tries to overcome these obstacles resorting to a decision which establishes some definite objective to reach.

On the one hand, this strategy constitutes a *de facto* renunciation of the reproduction of the exemplar in its entirety and, therefore, its replacement with a simplified model which privileges only one observation level. On the other hand, such a strategy implies the tacit admission that, if one wishes to integrate even only two observation levels, one must proceed, when it makes sense, to establish a third level, without assuming that it fully incorporates the previous two. Completely rebuilding an exemplar using a bottom-up strategy is, to sum up, a pure utopia.

Let us imagine designing the reproduction of the sub-system which controls and coordinates vision and touch in the human body, provided we know it well enough. Obviously all this will require setting up a model in which the relationship in question is central, assuming that the artificial devices for vision and touch are reliable enough. Such a model is quite plausible, thanks to the availability of electronic and computer technologies able to implement many types of very flexible complex algorithms. But the critical point is another. It would have to deal with the necessary adjustments we make in the artificial vision and artificial touch devices in order to make them compatible with a third artificial device, which adopts some region of our brain as exemplar.

In short, the true exemplar of such a design, would be the brain region itself and surely not the pure 'putting together' of the two partial artificial objects already available, which were designed and built as separate devices.

The coordination device, in such designs, is usually electronic or computer based, because, as we have already mentioned, these fields give researchers the most powerful working tools.

But the electronic or informational levels, which are certainly present and relevant in brain activity, do not exhaust its reality, since the brain is composed of other levels as well. Furthermore, the two levels in question will introduce, by inheritance, their own natures, which will also be imposed on the two partial artificial devices coordinated through some electronic circuit or computer program.

From this point onwards, the typical series of alterations of the sub-system, as compared to the natural one, will begin.

Thus, to cite a real case, the anthropomorphic robot Hadaly (one of the most advanced products in this field, built at Waseda University, Tokyo, in 1995), is structured in three sub-systems: a vision sub-system, an auditory sub-system and a motor one. But, beyond some minimal supervising algorithm of its behavior, Hadaly does at all not reproduce the performances we would expect from a man, with his capacity to coordinate, both on a reactive and a high level decisional basis, the events belonging to the three levels involved.

If one asks Hadaly where some institute of the university is located he would tell you and even points in the right direction with its hand, but if one approaches its hand with a lighter, Hadaly will not draw back or protest. Clearly, the introduction of an additional program in its computer, able to evaluate this kind of situation, would not solve the problem at all.

In scientific terms, i. e. looking at nature in such a way which does not aim to reproduce it but to know how things are, the logical aspect of the problem we are discussing is in some measure the same. As a typical case, let us cite the problem posed by a biologist referring to the relationships between vision and touch in the nautilus:

'Another fascinating problem is the relationship between visual and tactile learning. (...) Since the two systems overlap in the vertical lobe, maybe there is some kind of co-ordination between them. However, it has been demonstrated that the objects detected by sight are not recognized by touch' (Young, 1974).

In the second part of this book we will examine several examples of artificial devices within which one may guess many of the above problems and strategies.

It is worth noting the closed analogy between the constraints regarding the combination of observation levels in the field of the artificial and the cooperation among different scientific disciplines. Indeed, what we call 'interdisciplinarity', when it succeeds, consists of the setting up of a new observation level rather than a 'sum' or a generic 'synthesis' of two or more levels.

Biophysics or biochemistry are good examples of this kind Combining knowledge, lexicons and techniques of the disciplines involved, they both give rise to largely autonomous new sciences, no longer easily comparable in terms of knowledge, lexicons and techniques with the scientific fields they come from. Normally, from biophysics research we cannot expect new knowledge of a physical kind nor of a general biological type, but, rather, findings concerning those aspects of biology which are of an exquisitely physical relevance, such as the effects of radiation on living systems, the transmission of nervous pulses, muscle contraction.

When an interdisciplinary project does not succeed, one of the observation levels involved dominates the others. This may occur for several possible reasons, including the greater development of one of the disciplines compared to the others. It is a matter of a destiny which also seems to be true for some disciplines in the family of bionics, dominated, as we have said, by the temptation to reduce the most heterogeneous phenomena to a pure informational level.

Let us make some further observations regarding the possibility of obtaining cooperation from two or more artificial objects or processes in the same natural organism, for instance the human body. In principle, it seems that one would have little difficulty in putting two artificial objects, say A1 and A2, which had proven to be effective as stand alone devices, to work together. In fact, 'implantation' – to be distinguished from 'transplantation', which implies placing a natural organ in a body– of an artificial bone or an artificial heart in the same organism should not give rise to any problems. However, the implantation of an artificial duodenum and an artificial liver – though they are only available as experimental devices – would be a very different case, full of snares and degenerative possibilities due to the functional relationship that exists between these two organs in the human organism. We can surmise that the cooperation between an artificial liver, duodenum and pancreas would be even more complex.

Once again, the general problem is the variety of the observation levels involved and the arbitrariness which will govern the choice of the essential performances. Whatever model was used to pilot the design would only be able to select a basic level, relegating the others to the background. The resulting artificial device – a true artificial sub-system – would work well only until the assumed exemplar and its related essential performances, were determinant for the functional balance of the organism. Nevertheless, when some different performance was needed – an essential performance which would have required a different more complex model, the sub-system would start a process of uncontrollable degeneration of itself, the host organism or both.

In conclusion, in an artificial sub-system, consisting of more than one partial (local and separate) artificial device, the greater the *functional distance* among the organs chosen as exemplars, i. e. the more independent they are of one another, the greater the likelihood that the sub-system will function properly. But, this is clearly a rather ambiguous and uncertain condition if one considers of the deep interconnections which characterize reality in all its phenomenologies.

10   Emergency and Transfiguration: i.e. 'Something Always Occurs'

A famous Latin saying says that *senatores boni viri, senatus mala bestia* (senators are good men, but the Senate is a bad beast). In general terms, this means that the coexistence of single entities of a given kind, may give rise to a very different sort of whole which cannot be explained by or limited to the 'qualities' of its components considered individually.

Usually this phenomenon is defined by the term *emergence*, coined by G. H. Lewes in 1875. Often it is a matter of a simple change of observation levels. For instance, a mass of white and blacks microscopic granules will appear as such under a microscope, but, with the naked eye, the mass will appear gray. Nevertheless, in other cases, it cannot be denied that, though one remains at the same observation level, the 'sum' or the 'synthesis' of many objects or processes gives rise to something which goes beyond the features of the single parts. The principle of emergence constitutes a foundation on which the hopes of artificialists in several fields are based.

For instance, in 1987 Craig Reynolds, demonstrated that the coordinated flight of a flock of birds can be simulated without including any central coordinator in the model. Each simulated bird (or 'boid', as Reynolds called them), follows the following simple rules: avoid collision with other birds (*Collision Avoidance*), stay in step with other birds (*Velocity Matching*), try to stay as close as possible to other birds (*Flock Centering*). Lastly, each bird could only see its closest friend. The simulation – which was then adopted for successful movies – demonstrated that a flock organized in this way, was able to 'fly' on a computer monitor as a compact whole, avoiding obstacles to rejoin the flock just after them, exactly as it happens in observable reality.

Other models – always formal, i. e. simulated on a computer – in the research area which is defined as *Artificial Life* or, in short, ALife, are able to reproduce typical phenomena of living processes (self-reproduction, evolution, the struggle for survival, etc.) as they emerge from the coexistence or many single 'agents' (cells or 'cellular automata') and, therefore, from the relationships which arise among them.

Chris Langton defined ALife as

'(...) the field of research dedicated to the understanding of life through the attempt of abstracting the basic dynamic principles which stay at the basis of the biological phenomena, in order to recreate these dynamics in different supports – like computers – making them accessible to new manipulations and experimental tests' (Langton, 1989)

The spontaneous emergence, as a self-organization phenomenon, of intelligent behaviors or behaviors typical of life, sometimes depends on the achievement of a sort of 'critical mass'. This is what occurs in crowd phenomena or in nuclear chain reactions, or in all complex phenomena, in the evolutive differentiation of living systems and the emergence of intelligence in the human or animal brain.

Though the 'new types of manipulation and experimental tests', which Langton refers to, are virtually non-existent – if with 'experimental' we mean real tests carried out in concrete terms, and not just informational – the researches of ALife provide us with further food for reflection in the field of the artificial.

In addition to the hopes that it generates in the field of artificial intelligence or ALife researches – regarding the possibility that intelligence or life might suddenly and spontaneously emerge from their models – the principle of emergence may be useful for emphasizing a very simple truth, but, even in this case, permanently underestimated. We refer to the fact that, in the field of the concrete phenomena, but, in some ways – as we saw with the ALife, even in the informational area, though for other reasons – whatever system we build it invariably ends up by giving rise to something which goes beyond the objectives of our original design.

In this sense, emergence may be understood as an extension, or a particular case, of the inheritance principle. In other words, at a given observation level, the features of a system constituted by a certain number of components, may appear new compared to the features of the components as such because the latter properties belong to another observation level.

This happens because the relationships among the components may exhibit their own qualities, starting from the relationship itself, independently of the characters of the components in question.

Experience, even everyday experience, continuously shows us how true this is. Just consider the circumstances we might encounter in a laboratory or even a kitchen in which, on the basis of chemical or physical knowledge, we face reactions which produce new realities quite different from the simple 'ingredients' we have used. Likewise, in the sociological field, we know well how certain collective phenomena – panic, aggression, fanaticism, etc. – are due to psycho-social relationships which have no correspondence to individual motivations or attitudes: collective phenomena, clearly require relationships between individuals and this involves a shift of observation level.

Of course this is even more true in technological areas: the *black-out* which paralyzed New York, in the 1960s, as well as the many sudden break downs which strike machines or systems of many different kinds, generally belong to the same typology, i. e. to a class of events which emerges from a complexity and follows its an own logic, unpredictable and therefore uncontrolled by the single components.

The artificial is really nothing but a particular case of this typology, of course.

The point is, rather, that the quality of what may 'emerge' from an artificial device – i. e. the additional performances it may exhibit as compared to the essential performances included in the design – will simply not just consist miraculously of something similar to what the exemplar exhibits in nature.

The accurate reproduction of two natural organs – at their own observation levels with their own essential performances – which, in the human organism, also work together to regulate some other process, does not at all guarantee that the regulation process or performance will emerge automatically from their coexistence. This will occur if, and only if, a precise condition is satisfied: the expected performance must be a simple function (or consequence) of the essential performances reproduced by the two artificial organs.

To cite a rather general example, if we accurately reproduce some performance of sun light (e. g., arrays belonging to a well defined region of the spectrum which stretches from ultraviolet to infrared) and, simultaneously, a given quantity and quality of heat (dry, moist, airy, etc.), it is possible to obtain an 'artificial climate' suitable for the growth of a certain tree.

Indeed, we could state that an artificial climate will emerge from the combination of the two artificial processes in question, but it will be an emergence, as it were, guided by a design founded on sufficient knowledge to only generate that phenomenon and only that phenomenon. Actually, as far as artificial light is concerned, we have to accept the idea that it is always different from sun light, and thus

'(...) in the absence of bad weather, nocturnally migrating birds have been observed to be confused by artificial lights below them and are attracted at night to artificial lights when there is no moon combined with fog or mist at ground level in the area (...)'.

In short, birds are

'(...) drawn to light due to the differences in the properties of natural vs. artificial light' (Verheijen, F.J., 1958).

In short, all science consists of a description of phenomena of this kind, though, very often, we are not able to give some analytical explanation.

This is what happens with many drugs, therapies and physical products or processes we are able to reproduce suitably combining some objects or events though ignoring their genesis beyond the observation level at which we orient ourselves in order to obtain them. Among the more spectacular cases, just consider aspirin. We continue to discover more and more new effects of aspirin, including the recently discovered unimaginable reduction of the incidence of heart attacks in diabetic patients.

In our opinion, the most interesting aspect of all this is, the fact that we only know about researchers' success and do not consider their failures, which are certainly much more common. More precisely, what has 'emerged' from scientific or technical attempts, carried out throughout many centuries of research, which have failed – in the sense that they have not led to any interesting or useful phenomena? This question is very important if we consider how often it happens that experimental projects with specific objectives lead to 'emerging' phenomena of an altogether different kind which are considered to be so interesting that the research shifts its focus to these phenomena, just like the chance discovery of the semi-conductivity of doped silicon. In these cases, research fails in terms of fulfilling its original purpose, yielding to discoveries that lead to research regarding other purposes which do however produce very significant knowledge.

Even in all the other cases of failure – those in which the failure is not accompanied by unexpected discoveries – *something certainly does happen*. This is also true for the design of artificial devices: the failure of a design in this field does not mean that 'nothing happened', but that the performance which emerged had nothing to do with the performance considered to be essential in the natural reference model, or was only partially or weakly related to that function.

The principle of inheritance, on the other hand, draws our attention to the fact that, even in the case where we are able to satisfactorily reproduce the selected essential performance, an unpredictable series of other performances is, so to speak, lying in ambush when it is not present in a manifest way. This in turn means that the interactions between the artificial and natural world, including therefore interactions with human beings, will always depend on a much wider spectrum of performances than expected in the original project. In this framework, an artificial object or process which does not reproduce the selected essential performance is certainly a failure, but, after all, it is only a particular case of all the artificial objects and processes.

Naturally, this implies the possibility that, among the performances which emerge from an artificial object, some may be acknowledged however as typical of the model, even though they have not been designed intentionally. This possibility is obviously unthinkable and unpredictable. It can happen, but not thanks to the precision of the artificial objects and processes already reproduced since, as we have emphasized many times, it is very rare that a given performance is the simple, additive function of two other performances. On the contrary, it is more common that the performance in question, instead, is located at a different observation level than the ones adopted to design the two artificial objects and the relative essential performances. The essential performance of the two artificial objects and processes, in other words, can sometimes be adopted as the necessary conditions, which, however, are not enough for the emergence of a third natural performance.

However, what is certain is that, both for the principle of inheritance as such and for the principle of emergence – intended as an uneliminable derivative of any recombination of natural things – any artificial object or process can only be intrinsically intended to generate a *transfiguration* of the exemplar and its performances. Moreover, the transfiguration is even more amplified by the inevitable use of conventional technologies. As we have seen from the beginning of this work, conventional technologies can clearly be distinguished by their determined heterogeneity regarding the structure and dynamics of natural objects and processes which, in fact, such technologies intend to control and modify rather than imitate or reproduce.

It is worth noting that the transfiguration which we are referring to is not, in itself, negative. Very often artificial technology produces objects, processes or machines which exceed the performances of natural models which they get ideas from. This is the case for computers, machines which, among other things, reproduce the essential performance of logical or mathematical calculation, with a speed and precision clearly superior to man's. And this is also the case for more concrete artificial objects, such as those which are inspired by the structure and dynamics of fish.

However, transfiguration can only be avoided if, in selecting an observation level, we actually isolate our exemplar, and its essential performance, from all the other observation levels which characterize natural reality, thus defining a rather new, purified and stand-alone feature. This is, obviously, impossible working in the real world, while apparently it is easy when we simulate something on a computer, or when, in scientific theories, we resort to abstractions substituting the world with a simplified modeling of it. These are extremely useful undertakings and procedures which, in any case, are the only alternatives for human beings. However, they are also procedures which should not be confused *ipso facto* with discoveries of nature, even though at times they allow us to control it in well-defined circumstances.

Thus, in the case of Reynolds' *boids* and other ALife software creations, we find ourselves, in some way, in the same circumstance as the people who design programs capable of certain logical operations, such as deduction. If we enable the computer to make deductions and we tell it, on one hand, that 'all capital cities are big cities' and, on the other hand, that 'Paris is a capital city', it will be able to inform us that 'Paris is a big city', even though we never gave it such information directly. The fact is that the emergence of such information is essential to the deductive algorithm which controls the computer, but such an algorithm does not necessarily reproduce the human way of performing deductions. Men, probably, adopt a more biological and complex way than computers.

It is therefore legitimate to affirm that, even regarding simulations, although the selection of a single informational level obviously avoids transfigurations on other levels – for the simple reason that other levels, besides information level, do not exist –, eventually some form of transfiguration – such as processing speed and precision – still enters into play. This happens because the informational level, although it is fit to describe certain aspects of reality, once it is set up as a unique dimension, ends up itself by living its own life. The capability of a simulation model to generate behaviors which are comparable to the observable natural reality, at this point, is no different from the ability of mathematics to describe the natural world and therefore it can neither be defined as a discovery of the way in which nature functions intrinsically nor is it a reproduction of the world.

Of instance, today, according to Michael Tabor of the University of Arizona, Tucson, it is possible to mathematically describe phenomena which are not decipherable in other ways, such as the twining of tendrils which seemingly follows the same rules according to which a few spirals of bacterium or even our telephone cables twine themselves helicoidally. But this does not mean that it is able to reproduce a tendril. First of all, as the biologist Neil Mendelson, from the same university, commented, it is necessary

'(...) to describe exactly what happens to fibres in the real world' (Tabor, 1999).

11 Classification of the Artificial

On the basis of our reasoning, we can now propose a classification of the artificial objects and processes which explains their main features.

First of all, we have seen how artificialism has always led to two opposite kinds of activity depending on the *concrete* or *abstract* nature of the 'substance' by means of which the final product is designed and realized. Since man has always possessed and shown a distinct tendency to reproduce whatever surrounds him – and also whatever has a primary origin in himself, such as feelings, self-portrayals, etc. – it is not surprising that the entire history of man is intensely characterized by the invention and development of the most varied of technologies aimed precisely at the material expression, communication or reproduction of models of every kind.

Rock carvings and oral communication, the evolution of symbols and painting, the birth of music and poetry, up to the invention of printing and then radio and television, and the advent of informatic and telematic machines, are a few of the cornerstones which have marked man's effort to 'make common' individual and collective representations, as well as subjective and objective models and essential performances.

In spite of their incorporation in material objects and processes – rock or paper, sound, voice or colors, electromagnetic waves or electric signals – all forms of the communication or simulation of reality can be considered as abstract artificial objects and processes for the simple reason that their purpose is not to reproduce the model concretely and materially, but to reproduce our representation of it *as such.*

For example, when we decide to communicate the images and sensations we felt while admiring a certain flower, we obviously do not have any intention or ambition to reproduce that exemplar materially. In this case, each one of us – but also the poet or the painter, the writer or the computer simulation expert – only tries to share his own portrayal of the world, either by objective pretences ('now I will tell you how matters stand') or by openly subjective intentions ('now I will tell you how I see the world'). While scientific models, including even the most varied 'theories' belong to the first class, the artist's

'poetics', as well as each of our personal expressive styles, belong to the second.

A very different situation arises when someone decides to
attempt the material, concrete reproduction of a natural exemplar. In this case as well, as we have already emphasized,
we inevitably use representations (in the form of models) and,
therefore, we introduce a subjective and arbitrary dimension.
Nevertheless, our aim is to build something concrete which
hopefully everyone will agree on, so that anyone can recognize,
in the object or process that we realize, an object which is part
of any common experience: a flower or a heart, skin or rain,
but also, a little more ambiguously, intelligence or reasoning.

We then saw, or at least mentioned, how the artificialist's
strategy can predict an *analytical* reconstruction of the exemplar or a purely *aesthetic* reproduction.

From the very beginning the aesthetic reproduction privileges appearance – obviously verifiable on a well-defined observation level – of the model, independent of its structure.
Examples of this kind are found everywhere: in sculptures,
toys, architectural 'relief models', in various kinds of prosthesis, the wide field of so-called reconstructions, various types of
gadgets and in many industrial products known as 'imitations'. Usually, the essential performance of this kind of artificial objects is clearly commercially rather than scientifically
oriented, i. e. widely perceived in terms of economic 'demand'
which, being placed at the same observation level as the artificialist, grants its success.

On the contrary, an analytical artificial object pursues an
aim which, apart from its aesthetic appearance, coincides with
the exemplar's structure or, at least, with as much of its structure as is necessary to make the required essential performance possible. Therefore here structure does not mean the exemplar's form but, in a broad sense, its anatomy and physiology: thus structure is synonymous with all the correlated

parts, appropriately described on an observation level, in terms of material composition and functions.

The matter is not, however, altogether linear and simple. In some cases, the artificialist knows perfectly well that if he does not assign a structure to the artificial object which is similar to the exemplar's, the essential performance will not be achieved. In the case of certain artificial bones, as in many other bioengineering cases, it is obvious that the artificial device cannot have an arbitrary structure compared to the natural model, nor physical dimensions (mass, density, etc.) that make its implantation and functioning impracticable in the human organism.

Even for educational aims, bone subsystems have to respect an obvious essential performance: that of the overall dimension and that of all its parts. Industry products for educational use currently include several bone subsystems: the artificial skull of a fetus, an artificial human skull, male and female,

'(...) constructed from 14 individual parts, which can easily be dismantled and put back together by way of interconnecting plugs' (Somso, 2001)

as well artificial human skeleton, skeleton of the foot, skeleton of female pelvis, hand skeleton and even unmounted human skeleton.

In the case of the artificial eye – which is not a simple prosthesis but is intended to provide eyesight – it is even clearer how anatomy and physiology place precise structural and functional constraints on the designer, just as for the reproduction of other organs which, do not often show any exterior resemblance to the exemplar and are actually intended to be placed outside the organism.

If we then go on to the wide kingdom of natural objects and processes of an environmental kind (rain, snow, grass, landscape, etc.) which do not have a mere aesthetic or spectacular purpose but rather a realistic purpose in the most sci-entific sense of the word, it becomes clearer that the structural and dynamic knowledge of the exemplars cannot be ignored. As we have said, all this does not mean that the resulting ob-ject or process 'seems' aesthetically similar to the model. Rather, the essential performance of the model and the repro-duced essential performance will have to be similar.

In short, isomorphism – that is the equality or at least the formal analogy of the structures – is indispensable only when function is strictly connected to form and the artificialist in-tends to reproduce such a performance precisely in the same way. This happens when one attempts to reproduce the fingers of a hand using structures which, though very different from the natural ones, must bear some similarities with real fingers. This is due to the fact that the artificial device must present dimensions which are similar to the exemplar and allow for a few physical performances typical of natural fingers (prehen-sility, opposition of index finger-thumb, etc.). In other cases, such a computer designed for automatic translation, the exte-rior isomorphism obviously has no importance and nobody would expect the automatic translator to look like a brain or a human being.

Furthermore, what was previously said about anthropo-morphic robots becomes useful: the similarity, at a certain ob-servation level, between the model and artificial object or de-vice, is often important only in interfacing terms, i. e. to make the machine *human like* or at least *friendly* from the stand-point of the person using it. This fact, tacitly assumes that such interfaces, which actually increase the chance of deceit, could make it easier to interact with the artificial, avoiding people's difficulty in accepting the idea, often frightening, that the artificial has its own 'nature'.

Finally, we must keep in mind the fact that even among artificial objects, processes or machines which are concrete and analytical, very often the analysis of the exemplar's structure does not lead to the decision to reproduce it as such – either because of its complexity, or because of its intrinsic non reproducibility given existing conventional technologies – whereas it is considered possible to proceed with the direct reproduction of its essential performance. In these circumstances we can say that researchers and designers provide an applicable example of the principle of *functional equivalence.* This is a principle, which comes from biology but has also had applications in human sciences, according to which it is not rare, in nature or in social phenomena, to find that a certain function can be carried out by a different structure than the one originally intended for such a task. This is the case, in the technological field, of a digital circuit based on electromechanical parts, replaced by another based on integrated logic circuits, or, in the sociological field when a social institution tries to provide the socialization of orphan children by substituting the family.

In the field of the artificial we can assert that a high degree of functional equivalence is always present – and for this reason it is not useful in distinguishing a particular class of artificial objects – for, by definition, the artificial requires the use of different materials than those used by nature. Nevertheless, although the models which guide the artificialist on the one hand, and those that guide nature on the other, appear homologous at a certain observation level, it can always be shown that they will be different on all the other levels, with consequences that can be perceived in various aspects or qualities of the same essential performance.

In conclusion, the cross between the two dimensions which we have singled out – *abstract-concrete* and *analytical-aesthetic* – allows us to formulate a classification system which substantially covers all the possible cases of the artificial.

The artificial objects or processes which are mainly characterized by two extremes of the dimensions under consideration will be placed in the classes which are created in such a way, as shown in the table. For example, the functionality which is necessary for an artificial organ suggests that it be placed in the class of concrete-analytical artificial objects or processes, whereas a doll or a puppet will to be placed in the class of concrete-aesthetic artificial objects or processes, and so on.

The classification system we suggest does not only have the useful purpose of organizing a rather wide range of activities, objects, processes or machines otherwise assumed as undifferentiated. In reality, the main benefit of such a system – as in every other case of scientific taxonomy – is that it makes it possible to deduce of the general characteristics which an artificial object has according to its placement in one class or another. Therefore, we are dealing with a methodological instrument of remarkable potential importance, obviously provided that the individual classes are carefully recognized. In particular, it is plausible to claim that the constraints imposed by the selection of the observation level of the exemplar and essential performance, the problems which concern the reproduction of several performances, the principle of inheritance and the transfiguration processes, as characteristics of any artificial object or process, assume however different and specific features in the four classes which we have identified. This is a reality which the theory of the artificial has only just begun to investigate.

12 A Note About Automatisms

There is a region of the 'artificial kingdom', called the region of automatisms, which is found at the boundary between conventional technology and artificial technology. In this vast area of design – we can find examples dating back to ancient times, including the Egyptian technology of the pyramids. What dominates in this area is what we could define as the *principle of substitution,* the substitution of a technological device for actions once carried out by man.

A very clear illustrative case is the invention of the throttle valve for steam intake and exhaust in machines invented by Thomas Newcomen in 1812. The valve had to be opened and closed by a person, who had to pull the cord and release it at the right moment. According to various accounts, the person who was assigned to the valve, probably in order to avoid such a monotonous job, had the bright idea of tying the cord to the piston, so that the latter could pull the cord and release it.

As we know, the auto-regulation which we are talking about was later improved by James Watt using much more complex and efficient devices. However, the main problem we are interested in regards the actual artificial nature of these kinds of inventions. The problem concerns the conceptual difference between the reproduction of something and its substitution with something else. Automatisms almost always substitute human actions with a technological device which yields the same results (or better results). As such, they are not intended in any way to reproduce either the exemplar (normally man or one of his subsystems) or the natural performance which makes that action possible.

In short, we could say that the people who design automatisms, even though they normally concentrate their efforts on man's essential performance, are interested in its faithful reproduction only to the extent that the actions, which that essential performance makes possible, are effectively reproduced. For example, the device which automatically opens a department store door when clients enter and exit, could be defined as an 'artificial porter' only by isolating the effect of the essential performance (opening and closing the door) but certainly not by the designer's effort to reproduce human vision of moving objects.

Likewise, *brain wheels* (gears moved by *cammes*) which made the mechanical programming of machine tools possible, during the last century, had the purpose of automating machine tool's work by substituting man's action but without any reproductive aim of the human mental performances which make those actions possible. The same can be said for a thermostat which activates or shuts down a radiator or refrigerator, the automatic shifting of speed ratios in a car, the stabilization circuits for the operation of a television set and, generally, all auto-regulation devices based on *feed-back*.

A computer, programmed to administrate a firm's accounts, can be included in this category since the modalities it uses to perform calculations have nothing to do with those adopted by man, just as the structure on which it is based has nothing in common with the human brain. On the other hand, in the field of artificial intelligence, programming a computer so that it produces stories or translates from one language to another means starting from a human model and taking into close consideration the essential performances which are involved, while trying to reproduce their dynamics.

However, this does not mean that, translating machines, based on models which are deliberately independent of the structure of our mind, are not only conceivable, but actually more efficient than the ones inspired by the first models of artificial intelligence, which insisted on the discovery of the human way of performing a translation in order to reproduce it.

Both in the case of artificial in the strict sense of the word and in the case of automatisms, the final resulting action which results from the device can be 'deceiving' to whoever interacts with it: a correct commercial invoice does not allow us to know if it was made out by a computer or a human being just as a heart does not know if the impulses for its rhythm come to it from the natural organism or from an electronic *pacemaker*. Nevertheless, the distinction between artificial and automatism is of great importance. In fact, whereas an artificial object or process is focused on natural performances which, often, can lead to different actions, automatism directly substitutes the action and therefore could be defined as pragmatic artificial. In other words, an automatism includes an action in itself, rather than the cause that produces it.

It is clear that the reproduction of a hand muscle and its natural physical essential performance, which allows a patient to work in the real world, within certain limits, as he pleases, is a much different thing than the substitution of human actions in a mechanical automatism which accepts paper money in a change machine. The same is true, in short, for the analytical reproduction of any natural exemplar in which the essential performance is considered more important than a specific action which it makes possible. In this sense, a robot for generic uses – although today we are still far from being able to give it a sufficiently broad generic intelligence – constitutes a clear example of an artificial object whereas an automatic cart which is able to move skillfully inside a factory only constitutes an example of automatism, albeit advanced automatism.

One of the characteristics, though not the most important one, which differentiates the artificial in the strict sense of the word from automatism, i. e., reproduction from substitution, is, therefore, the degree of flexibility involved. The greater it is, the more legitimate it is to speak of the artificial, since, in order to give a device greater flexibility we inevitably must go from final action models to more and more generalized models of natural performances which, for example in human behavior, make that action feasible.

It is under these conditions that, in artificial intelligence, we have experienced and are experiencing the problem of the shift from *micro worlds* to more general models. Micro worlds are local models of the world which, for example, describe the typical situation found in a restaurant with waiters, customers, chefs, tables, the coming and going of people, etc. By communicating this micro world to a computer, by means of appropriate programming, it has been shown that you can get adequately intelligent answers from the machine to questions that regard rather broad classes of events that can happen in a restaurant. The same can be accomplished, naturally, in other micro worlds, for example, supermarkets. However, it has also been proven, as we have already noted regarding the problem of the synthesis of observation levels, that the intelligence tested in one micro world is not interchangeable with the intelligence produced in another one, except if we build a new, not so 'micro', model of the world. Without reproducing generic intelligence as a natural performance – and therefore also so-called 'common sense' – we would thus soon be able to reach dimensions of the problem that are definitely beyond our capabilities.

However, the most important characteristic which distinguishes the artificial from automatism is the former's close necessary link with nature. In other words, the artificial in the strict sense of the word, once again confirms its heavy dependence on nature – even though it subsequently tends to transfigure nature –, whereas automatisms possess a much more marked technological-conventional tendency from the very beginning.

Although we are dealing with a distinction which can turn out to be very subtle – since any artificial object or process, as we have emphasized many times, requires conventional technology – it allows us once again to recognize the two basic aspects of technology, from a motivational and intentional standpoint.

The pragmatic aim of technology, the domination of nature, is in fact pursued both by means of pure inventions and inventions which try to produce the same effects or actions as nature without any concern for the similarity of the structures or the processes in question to those found in nature.

The reproductive aim of artificial technology, on the other hand, can only be pursued using strategies which give primary importance to the exemplar and its performances as such, regardless of their effects or actions. In short, in something that is truly artificial, there is always a high degree of homology (similar structures) or analogy (similar functions or relationships) with the exemplar and its performance, while automatism does not take this into consideration. For example, the compatibility of the materials and mechanisms of an artificial limb with the rest of the organism – its homology or at least its analogy with the natural model and its essential performance – becomes an integral part of such a project; on the other hand, if the pragmatic aim is only to allow the patient to move a bit, any other external mechanical device, almost completely neglecting the anatomy and physiology of the limb in question, will suffice.

In this sense, the technology of the artificial can be considered an activity that is very closely related to scientific research, because, in order to advance, it needs knowledge and models of the exemplars and their performances while, at the same time, it can contribute, at least hypothetically, to the advancement of this knowledge. On the contrary, the technology of automatisms, while sharing artificial technology's goal of substituting nature or, more often, human actions, cannot contribute in any way to the scientific knowledge of the exemplars or their performances, precisely because in order to advance, it does not consider their intrinsic features and behavior.

Finally, whereas artificial technology in the strict sense of the word shows the human desire to recreate nature – an undertaking which obviously includes the pretence of having an excellent knowledge of nature – conventional technology and automation technology are evidence of man's desire and capacity to control things and events in the natural world, by getting or without getting ideas from their way of being in nature.

To sum up, every technological device which can actually be defined as artificial acts as a substitute for something natural, while it is not true that a device which aims to substitute something natural is, in itself, something which tries to reproduce natural things in the same way artificialists do.

We must keep in mind, in fact, that many products of artificial technology have nothing in common with the world of automatisms, although they involve the substitution of a natural object or process with something technological. For example, the domes in contemporary Japanese architecture (huge constructions that reproduce, among other things, alpine or tropical landscapes in which people can go on vacations) are certainly artificial products and, as such, substitute their own natural exemplars. However, they do not constitute the automation of anything. The same can be said of various artificial organs which are implanted into the human organism, and substitute one of its performances as well as many drugs which are often actually developed as imitations of natural substances.

Classification of the Artificial

	concrete artificial (material devices or processes)	abstract artificial (informational devices or processes)
analytical artificial (reproduction of structures)	**A** • organs • cells and tissues, • robots • Virtual Reality interfaced with the real world • miscellaneous (e.g., diamonds, grass, horizon, etc.)	**B** • AI (Artificial Intelligence) • ANN (Artificial Neural Networks), • ALife (Artificial Life) • GA (Genetic Algorithms) • (...)
aesthetic artificial (reproduction of appearance)	**C** • sculpture • architecture • imitation gadgets • reconstructions, • (...)	**D** • drawing • maps • figurative arts • simulated graphs, • descriptive Virtual Reality or Virtual Environment

2 The Reality of the Artificial

1 The Bionic Man

Artificialists' greatest aspiration, needless to say, is the reproduction of man. Marvin Minsky, among others, maintains that exploiting robot and artificial intelligence technology in order to repair our body,

'will be making ourselves into machines' (Minsky, 1994)

reducing the question of our identity to a simple, slow and pleasant evolution. Others, like Hans Moravec (1989), Ray Kurzweil (2000) and Bill Joy (2000), predict that robots and AI programs will even supersede humans, menacing our species or, at least, making our nature irrelevant. In light of what has been said in the previous sections, it is very difficult to agree with such positions, many of which have already been exposed, by the way, in the central chapters of *Erewhon*, by S. Butler. Written in 1872, the book deals with the consciousness, and supremacy of machines, and related fears (Butler, 1988).

For instance, Butler asked

'Are we not ourselves creating our successors in the supremacy
of the earth? Daily adding to the beauty and delicacy of their organi-
sation, daily giving them greater skill and supplying more and more
of that self-regulating, self-acting power which will be better than
any intellect?' (Butler, 1872)

However, as we have seen in previous sections, such an
ambition, besides being unfeasible, also has several methodol-
ogically vague aspects which end up becoming real obstacles,
in principle, in the realization of the task. The main point
seems to regard the observation level: what does the word
'man' mean without any adjectives – such as anatomical,
physiological, psychic, mental, social, etc. – that define the dimension adopted for reproduction?

We already know that this is a problem we encounter in
any artificialistic activity for, when an observation level is not
explicitly established, the model remains undefined, without
any specification of the 'profile' we intend to select in order to
observe and, subsequently, reproduce it. However, we also
know that the choice of an observation level is an inevitable
fact, which is part of our natural habits, closely connected to
our limits in interacting with the world.

This means that, even without any explicit indication by
the artificialist, the observation level that he chooses, even un-
consciously, will clearly appear in his concrete activity, i. e., in
the aspects of the exemplar that he actually tries to reproduce.
The adjective that is missing in the expression 'reproduction of
man' (or of any other natural exemplar) is filled in by the reality of the designer's actions.

Indeed, in the history of civilization, the real or imaginary
cases of the reproduction of man are practically all oriented
towards the implicit assumption of observation levels which we
could call 'everyday', namely observation levels we adopt in our
daily life through our common perception of things. Actually,

this is the most ambitious choice, because an 'everyday' or generic man – unlike a man who plays a specific role, such as a
physician or diver, policeman or customer – is a man who can
think and act in many different, constantly changing ways according to heterogeneous situations. He has consciousness
and common sense abilities as well as many other aspects.
This is why contemporary robots, on the contrary, are usually
designed on the basis of a much more narrow definition of
man, almost always determined by some specific function or
performance they have to accomplish.

In ancient China, as Joseph Needham recalls, it is said
that an inventor, Yen Shih, offered the king, who was visiting
his town, an 'automaton' with features so real that the king
himself had a hard time to understanding what kind of gift it
was since, next to Shih who was speaking to him about a gift,
he only saw a man who he thought was the technician's guide.
Indeed, as Needham reports,

'(...) anybody would have mistaken him for a human being (...)'
(Needham, 1975)

After learning that the man was, in reality, the gift in
question and that he was an artificial realization, the king was
astounded. The automaton, indeed was able to move, sing,
look, etc. like any human being. Yen Shih even ran the risk of
being condemned to death when the automaton began to court
one of the concubines who accompanied the king. At this
point, Shih showed him the 'pieces' that made up the automaton and the king was able to verify that it was a creature made
of leather, wood, glue and lacquer, all painted in white, black,
red and blue. Examining it more closely inside, the king recognized the liver, heart, lungs, spleen, kidneys, stomach and in-
testine and, over these organs, muscles and bones as well as
the various limbs with their joints, skin, teeth and hair, all artificial of course.

By putting all the pieces together again, the automaton regained its initial appearance and behavior. By removing a certain organ, he lost the function which depended on it. The king was enthusiastic and thought that this might be proof that human ability could compete with the ability of the great Author of Nature.

In Greece, as we recalled in the first chapter, mythology speaks of numerous 'living' creatures. We can also mention the Delphic oracles, which spoke through the wind or Talos, an automaton which Efesto built to watch over Crete.

There is also the Hebrew legend of Golem, a man of clay who was given life through letters of the 'alphabet of 221 doors', on all of his organs. In this case, we have a very interesting *sui generis* reproduction, since the words are given the power of life and death over Golem: indeed, on his forehead the word *emet* (which means truth) had to be affixed, whereas in order to destroy him it was only necessary to eliminate the first letter, to get the word *met* (which means 'death').

Unlike the concrete artificial imagined in the Chinese anecdote, Golem is concrete in appearance but significant above all for his 'shift of state', from life to death. We find ourselves before a sort of a 'programming' of the automaton, obtained through of appropriate language and the right 'algorithm'.

In the Renaissance, the doctrinaire tradition which dates back to the mythical Ermete Trismegisto, was revived by Marsilio Ficino and Pico della Mirandola. Together with alchemy with its ability to transform one substance into another, this tradition favored attempts which aimed to reproduce life using mechanics. As Bruce Mazlish writes,

'In the hermetic tradition of the Renaissance, the ancient charm inspired by automatons acquired new impulse. Magic and mechanics united intimately, and an air of amazement and fear rested on statues and angels which appeared on earth and in the air: are they real and living beings or not? Are mechanics, in turn, human beings, given that they are able to give life to whatever they build by imitating, in such a way, their own Creator?' (Mazhlish, 1995)

Through the centuries that followed, which saw Descartes' rationalistic lesson and Bacon's empiricism, up until today, the production of automatons or machines capable of reproducing a certain aspect of living beings, above all movement, has become a less mysterious endeavors mostly oriented towards science.

In the eighteenth century, Jacques Vaucanson appears to have been one of the more conscious designers of this new tradition. Even though, as Mario G. Losano recalls, he was suspected of using tricks, his main purpose was not only to astonish, but, at least to a certain extent, to faithfully reproduce what happens in nature. For example, regarding the digestive system of his artificial duck, Vaucanson claimed that

'(...) the food is digested in the same way as in real animals, by decomposition and not by trituration, as some Physicists maintain.' (Losano, 1990)

Note the following design by Vaucanson – described by Gian Paolo Ceserani – presented by the French technician at the Academy of Lyon in 1741 for the

'construction of an automaton which will imitate in its movements all life functions, blood circulation, respiration, digestion, muscle, tendon and nerve movements and so on. The author believes, by means of this automaton, that he will be able to experiment on animal functions and make inferences in order to acquire knowledge on the different phases of human health, in order to find a remedy for its illnesses. This ingenious machine, which will represent a human body, will be used in the end for a demonstration in an anatomy course' (Ceserani, 1969).

The definition of ALife's goals, described by Christopher Langton in 1992 – see the chapter 'Emergency and transfiguration, i.e.: something always occurs' – sounds significantly similar, particularly concerning the possibility, indicated by Langton, of

'(...) reproducing the dynamics of life in other physical supports and thus making them accessible to new types of experimental manipulation and control' (Langton, 1992).

Though both Vaucanson and Langton are aware that they are operating on materials which are very different from those provided by nature for its living products, they still believe that it is possible to acquire knowledge on real life. Nevertheless, due to its formal stuff,

'Computational modeling (virtual life) can capture the formal principles of life, perhaps predict and explain it completely, but it can no more be alive than a virtual forest fire can be hot.' (Harnad, 1994)

Likewise, the choice of a single observation level – anatomical, physiological or informational – regardless of the precision of the reproduction of structures and processes which are made accessible on that level, does not prevent artificialists from considering their designs as reproductions of natural exemplars, or from carrying out studies on these designs as if they were natural reality itself. The complexity of the natural exemplar, which cannot be reduced to one dimension or level, does not prevent both Vaucanson and, in more sophisticated terms but inevitably equally limited, ALife researchers, from following a more ingenuous principle according to which the *visibility of movement* (of a duck's muscles or of a 'colony of cells' on a computer monitor) is proof that reproduction has taken place.

However, it is interesting to note that the mechanical observation level, typical of the eighteenth century and consistent with the development of anatomy and physiology in that era, and the informational observation level, typical of today's development of informational theories and technologies, are quickly replaced by more pragmatic observation levels when it comes to building concrete artificial devices, such as industrial or military robots, or robots of any other kind.

In these cases, the 'reproduction of man' very soon becomes the reproduction of some of his well-defined essential performance reproduced consciously without any noteworthy reference to what occurs in human beings, above all when a superior function is involved, such as recognition, control abilities, typical problem solving, calculation, etc. In short, in the field of robotics, it is fully acceptable, to use an expression introduced in the first part of this work, i. e. that the artificial involves a transfiguration of the model and its performances. In fact, robotics often specifically aims to transfigure in order to pursue its own pragmatic goals. For example, a robot, or bionic man, can be given the (certainly non human) ability to make calculations or comparisons and rational decisions in environments full of gas or right in the middle of a battle; the power to rotate its wrist 360 degrees or to focus its hearing on a few frequencies and exclude others; the ability to find in its memory, with 'inhuman' speed, geometrical shapes or very complex data configurations.

Naturally, artificial technology can be very useful for scientific or educational purposes as well, as Vaucanson had already sensed. Today, we use extremely useful artificial 'bodies' to test surgical techniques or seat belts, ergonomical models and various other types of devices.

However, the bionic man remains an unreached – and intrinsically unreachable – dream if, by bionic man, we mean even only a physical reproduction of the human organism, which can operated with or on as if it were a living 'general purpose' organism. With a few precautions, this reproduction may be possible, as we have said above, for specific purposes, i. e., where the essential performance of the human body, considered essential in some specific situation, is persuasively

and efficiently isolatable from the organic context of its whole
and sufficiently independent of materials.

For example, so-called based-based robotics certainly
seems to be very promising approach which aims to give robots a more flexible behavioral intelligence which is more similar to human intelligence in certain categories of real situations. In these designs, the robot must be able to select the
best based among those available, on the basis of assumptions
and evaluations of the situation in which it must act.

Such a purpose will nevertheless be pursued with a great
deal of pragmatism, namely compelling the machine will be
made to carry out the desired behaviors by means of software
and hardware devices which constitute the 'tricks' and 'illu-
sions' of the usual artificialism. Thus, the based-based strategy can, for example, be combined with the object-oriented
strategy and therefore, as R. Bischoff, V. Graefe, and K. P.
Wershofen write,

'(...) the robot will only recognize the objects which are important in order to carry out his operations correctly. Consistent with
the object-oriented vision concept, the recognition is greatly reduced
to a knowledge-based verification of objects or characteristics which
are expected to be seen in the various situations'. (Bischoff et al.,
1996).

Even robots controlled by neural systems (ANS, *Artificial
Neural Systems*) are making their theoretical or prototypical
appearance, making the best use of the speed and flexibility of
neural systems while adapting to the configurations of the environment and its alterations. With these strategies we may be
able to design 'artificial creatures': systems that are able to
perceive the events of the outside world with great precision
and act accordingly by means of real-time planning of the actions to be carried out; these 'artificial creatures' will also be
able to learn from experience and possess some kind of motivations, purposes, etc.

The main problem lies in coordinating all these and other performances. The solution must include some human, prejudicial decisions on the observation level which is most appropriate for a given situation and certainly not the tendency to set up overall models whose rules are not known for even the simplest animal's brain.

2   The Universe Under the Microscope

One of the most important conclusions that we reach using the theory of the artificial is that, given any exemplar, its faithful and overall reproduction is hindered, first of all, by the impossibility of describing it fully and faithfully.

The insurmountable obstacle, as we have seen, is a result of the selective nature of our observation: we cannot consider and describe any kind of object by simultaneously taking into account all the possible observation levels. However, this is precisely what we must do for a complete reproduction of the exemplar, either by adopting materials that are different from natural ones (the artificial), or, ideally, using the same materials (replication).

This problem is generally easily understood if, as exemplars, we select objects or processes from the natural world which we believe to be not only very complex, but also very closely linked to their 'background': man, but also any kind of animal, a flower, a pond, a plant, life, evolution, etc. Having to describe one of these realities, we immediately realize the extreme arbitrariness of the operation, which will always depend on the selected observation level. We cannot seriously consider a strategy which aims to grasp them completely and therefore simultaneously on all possible levels.

On the other hand, when we examine the possibility of reproducing exemplars which seem to be 'parts' or subsystems of whole objects (like the ones mentioned above), it seems to be a less difficult problem. Indeed, we cannot deny that the prospect of reproducing man's limb or skin – although it is certainly not an amateurish task – is not as difficult as the reproduction of man in his entirety.

This is easily explained, since many subsystems have already been defined and described for centuries, not only as objects of daily experience, but also anatomically and physiologically, i. e., by acquired observation levels which have become an integral part of common sense and general perception. Likewise, the essential performance of an organism's subsystems appears to be clear and at times even 'obvious', above all, at a macroscopic physiological observation level: thus muscles are levers, the heart is a pump, the liver is a filter, etc. On one hand, this explains why artificial intelligence researchers and designers grapple with a thousand controversies, difficulties and mysteries. In reality, in our minds, there are no reinforced observation levels or confirmed knowledge regarding its various functions. Thus, designers find themselves in the same situation that bioengineers would have found themselves before acquiring knowledge, for example, on the physiology of the bloodstream or the function of cardiac valves, discovered in the XVII century by William Harvey.

On the other hand, what we have emphasized makes it easy to understand why, in the field of the concrete artificial, designers advance with considerable speed and success.

Nevertheless, we must bear in mind that, both those who work in uncertain, complex and abstract artificialistic fields – such as artificial intelligence or ALife researchers – and those who work in more established concrete fields – such as bioengineering – have never been dissuaded or discouraged in their work by the above-mentioned difficulties nor do they have any real reason to be. Indeed, in order to design something artificial, the artificialist must only possess, first of all, a representation or, better, a shared model of the exemplar and its essential performance and, secondly, the materials and technical knowledge necessary for its reproduction.

There is naturally a crucial difference between a reproduction based on representations which, in turn, refer to scientifically confirmed information and models – and also, implicitly, to some preferential observation level – and a reproduction based on hypothetical or very subjective representations – perhaps without any sure indications with regards to observation. This difference lies in the fact that, in the latter case, since there is no common observation level, the artificial object's based will not be considered persuasive in spite of its possible effectiveness at a certain level For example, the reproduction of human intelligence according to a model which describes intelligence as the formal ability to follow rules, could lead to the construction of a very useful, but unconvincing device, for it is clear that a reliable model of human intelligence should have much more than the ability to follow pure formal rules at its core. All this regardless of the fact that the model may turn out to be appropriate at another time and through other channels. On the contrary, a common observation level, together with the existence of confirmed knowledge, make the field in which the artificialist works more certain, and the acceptance or rejection of the artificial object which he produces easier.

This long preamble has the sole purpose of confirming the very delicate role of our choices, before passing on to what we could call the most emblematic and dynamic field of research concerning the artificial, namely the field of artificial organs.

In this field, the work of bioengineers is becoming more and more fascinating and efficient but is often accompanied by problems which seem to increase precisely as knowledge advances. Some of the greatest difficulties emerge precisely when researchers are forced to consider the observation level, and

thus the essential performance of the exemplar, which are unexpected at the beginning.

This problem, as regards several types of biomaterials
needed for artificial organs, has been summarized in 1995 in
the following formula by Tirrell and Hoffman, two researchers
from the Minnesota Biomedical Engineering Center

'(...) if we wish to predispose a material which has the characteristics of soft compound biomaterials, we must study the interactions which are involved at all levels: between molecules, up to the
cells, up to the macroscopic characteristics of the tissues involved
(for ex., cartilage, tendons, skin, editor's note)' (Tirrell 1995, in Hoffman 1995).

3   The Boundary Between Illusion and Compatibility

Every organized system defends its boundaries in order to
safeguard its identity, and biological systems are masters at
this. In this sense we can assert that the theme of biocompatibility and biofunctionality sums up well a part of our comments regarding the difficult and perhaps, beyond certain limits, prohibitive ambition of reproducing natural objects – for
example human beings – capable of engaging in normal relationships with the rest of the environment or organism.

Biocompatibility can be defined as a property of surfaces.
It is here, in fact, that an organism recognizes, and thus accepts, or does not accept, any substance it comes in contact
with. In other words, the interaction between an artificial organ and the biological environment of the organism, begins
precisely at the surface of tissues and cells.

According to several researchers from North Carolina
University who are involved in the reproduction of the esophagus, as in all cases where a natural organ is replaced by an artificial one, the end result is that, complications which can be

fatal often develop. Complications associated with the implantation of an artificial esophagus include: anastomotic infiltration, infection, erosion, stenosis and displacement of the prosthesis. Infiltration and infection develop when there are biocompatibility problems between the prosthesis and the host organism.

Since even the use of bioinactive materials – which should not interact significantly with the organism – has proven to cause complications, researchers are now working on bioactive materials which, however, do not offer the same advantages as inactive materials. Hence, a hybrid artificial esophagus has been designed. It is partly based on traditional, conventional materials and technologies and partly based on biomaterials, or special conventional materials which display sufficient compatibility with the organism. The device consists of an internal tube and an external tube made of a sponge of hygienic collagen. The collagen has been applied to promote regeneration of the tissue, especially the epithelium.

The silicone tube is thus soon covered with natural tissue and, once regeneration has taken place, it can be removed because the cells do not adhere to the artificial esophagus, since silicone is one of the most bioinactive materials. The model and its essential performance are therefore only reproduced artificially *pro tempore*, protecting the organism's boundaries and leading it, at the same time, to self-reproduce the organ towards which there will no longer be any immune defense response (Takimoto, 2000).

A similar strategy is also followed for artificial bones. As an expert, Jillian E. Cooke, wrote,

'A bone is a fascinating 'nanocompound' (composed of extremely small structures, editor's note) object: it is hard and strong at the same time. Its properties have proven to be extremely difficult to reproduce using conventional materials. Recently, researchers have synthesized a ceramic-organic nanocompound using strategies which imitate the way in which the organism synthesizes a natural bone' (Cooke, 2000)

Within the framework of a hybrid strategy, in fact, a 'matrix' is adopted – a biomaterial structure allowing cell growth – made of a compound called apatiteorgan, which acts as a 'root' for cell action and bone growth. *In vivo* experiments conducted on rabbits at the University of Illinois have demonstrated the reabsorption of the artificial interface, followed by regenerative processes of the femur.

This is a technology which still has many uncertain aspects, but which seems promising. According to the Laboratory for Surface Science and Technology (LSST) in Zurich, the two main aims in the study of the based of biomaterials on surfaces consist, first of all, in a deeper knowledge of the superficial processes which come into play when a biomaterial comes into contact with an *in vitro* biological environment. Here the problems concern the alterations of the composition and structure of oxide films, the adsorption (the process of accumulation of molecules on an interface, editor's note) of biomolecules, and the interaction with cells.

The second aim consists in modifying the biomaterial surface in order to obtain specific properties and improve the biocompatibility and biological functionality of the material that constitutes the artificial device. Regarding the materials used, research is now concentrating on titanium and its alloys on the market.

The physical boundaries between artificial and natural in the field of artificial organs are so important that the International Standard Organization (ISO) and the American Federal Drug Administration (FDA), establish very clear typologies in order to provide a useful classification for biological compatibility tests. For example, ISO establishes three basic categories: devices which are placed on the surface, external devices which communicate with the organism and implantations.

The case of artificial skin is, obviously, especially crucial in this framework since it simultaneously concerns the first and third category of devices. Today, various products exist in this area. Two devices have been approved by the FDA: Integra – Artificial Skin Dermal Regeneration Template – and Original BioBrane. Integra is a two-layer membrane, the first layer comes from bovine collagen and, together, from a substance (*glycosaminoglycan*) manufactured so that it presents a regulated porosity and a well-defined degradation rate.

The temporary epidermal substitute is made of a synthetic (silicone) polymer and acts to control the loss of humour caused by a burn. The infiltration of fibroblasts, macrophages, lymphocytes and capillaries which come from the bed of the burn are generated through the first layer. As the healing progresses, the fibroblasts deposit an endogenous matrix of collagen and the layer of artificial skin is degradated.

In a subsequent stage, after an adequate vascularization of the dermal layer – and with tissue taken from the same patient (autograft) – the silicone layer is removed and the thin layer of epidermal autograft is placed on the 'neodermis'. The cells of the autograft will grow and form the corneas layer, closing the burn and making the functions of the dermis and epidermis once again possible.

However, the strategies described above are not only typical of the world of bioengineering. Indeed, the concept of biocompatibility constitutes a particular case in a wider field which concerns the absolutely general problem of matching the artificial and natural. This matching is based on the need to control the transfigurations that the artificial could set off because of the principle of inheritance and the choices it comes from (observation level, exemplar and essential performance).

An insightful American writer, Ken Sanes, has dealt with the strategies adopted by the designers and creators of artificial environments or landscapes (*artificial naturescapes*) created to 'deceive' people, for example, visitors at an advanced technology zoo. Naturally their first task is to carefully observe, directly and indirectly, the environment in which the exemplars of their artificialization task are found.

As part of the general project of Lied Jungle, in Omaha, Nebraska, many specialists from the construction companies involved, visited Costa Rica where they thoroughly studied the characteristics of the trees necessary to reproduce the forest. The main operation consisted in obtaining the external shape of the bark by means of latex layers. After the latex had dried up, they had the tree-trunk's imprint.

Once transferred to Lied Jungle, the imprints were pressed on appropriately shaped concrete columns measuring 15 to 24 meters, thus obtaining a formal 'replica' of the bark. The same technique was used to reproduce the rocks, which were touched up by hand, as well as the plants, using colors which were as close to the natural ones as possible.

When this stage was completed, a few natural plants were mixed with the artificial ones and the result is that the visitors, while admiring such an organized landscape, lose their ability to distinguish natural from artificial just from the 'surface'.

In Jungle World also, another landscape installation, the visitor, by standing in a certain position, can look at two layers of leaves: the leaves of the first layer are made of polyester, while the second ones are rubber plant leaves, and it is impossible to distinguish them.

We now move on to another observation level, the acoustic level audible to man. In Jungle World loudspeakers hidden in the trees continuously emit the chorusing of insects and birds, recorded in a Thai forest, which is mixed with the sounds produced by the animals in cages.

The strategic requirement of invisibility, which protects the natural on one hand (the audience which must be deceived) and artificial on the other hand (the illusory devices), is thus met according to a logic which is perfectly analogous to the one that bioengineers follow in their task of separating organisms and artificial devices.

Indeed, up to this point, Sanes comments,

'(...) we have only begun to scratch the surface of illusion. If we deepen our investigation, we will discover that these "immersion landscapes" are also based on secret simulations, that is, those that use a partial invisibility, cover-ups and distracting events in order to hide characteristics which could interfere with the illusion. As in every good trick of magic, art is what the audience cannot see and what it does not know' (Sanes, 1998)

Rightly, Sanes emphasizes that,

'(...) with a few variations, these characteristics can be found in all "worlds" invented by man and which become part of popular culture: from theme parks to films to virtual reality machines' (ibidem).

Furthermore, to avoid contact between organic elements (zoo animals) which would be, so to say, 'rejected' by the host system (the audience), while allowing the audience to enjoy the view in a realistic environment (the true essential performance of the whole installation), a whole series of devices are used. They include artificial rocks and islands on which the animals can be observed and, at the same time, isolated; thin electrical wires and even Vaseline on tree trunks and branches to prevent the monkeys from climbing up the trees.

In short, Sanes concludes,

'(...) it is obvious that in these installations we witness a drama which is very different from the visitors' demand to observe animals in their natural habitat. (...) We can say that artificial tropical forests are a kind of theatre with a cast of prisoners or animal actors who continually try to disappear from the scene' (ibidem).

Here we also have a considerable contribution which comes, as we might expect, from conventional or artificial technology. Air conditioning ducts, drainage and irrigation lines, land supports, feeders and everything necessary to facilitate the sensation of being in a tropical forest are well-hidden in rocks and trees or behind other perceptive obstacles.

Understandably, not everything is hidden and many things are completely excluded, such as the sights, sounds and weather which should surround the environment. Other things however – such as side effects which inevitably accompany artificial devices or infections which affect artificial organs – occur unexpectedly and create difficulties for the installation. The intrusion of some forms of unauthorized wildlife is a particularly common problem. These intruders include mice and cockroaches, which enjoy the artificial forest habitats, with their ideal climate and abundance of food and places to hide.

Hence, caretakers are constantly at work not only to keep a few natural organisms alive, i. e., the zoo animals, but also to eliminate other forms of life which get into in the crevices of the constructions.

In short, the caretakers act as highly selective drugs or antibodies to 'regulate' the processes which are considered essential. Here the problem of compatibility takes on the form of man's careful and intentional selection of the animal species that will be authorized to survive in the zoo, in order to prevent the development of a completely natural food chain, but one which is harmful to the installation's purposes. Thus, toucans have been removed from Lied Jungle because their behavior, which includes destroying, taking away other birds' babies, and eating various kinds of frogs and lizards, is considered unacceptable.

As in many bioengineering cases and, in the end, in all artificial objects and processes, 'deception' and illusion rely on expensive and complicated monitoring and control systems, which aim to hide and protect something while allowing something else to be activated for the achievement of the essential performance.

The problem is even more real in cases in which man tries to get nature to reproduce itself setting up the right conditions, partially artificial and partially conventional, which nature needs to function. While this is becoming a real possibility, starting from the deepest structure of biological systems, i. e. the DNA, it is much more difficult where the open, natural environment is concerned. These kinds of difficulties were encountered in Texas, where man has tried to recreate several wetlands for environmental purposes. According the *Texas Water Resources*

'Some studies show that as many as half of all created wetlands fail to achieve desired goals. Concerns revolve around such issues as the complexity of reproducing natural systems, the difficulty of measuring the success of man-made wetlands, the ability to mimic wetland functions such as flood control or water quality improvement, the extent that aquatic life will utilize the sites, and long-term success' (TWR, 1992).

Indeed, in these situations, the task of establishing the boundaries of the exemplar is very difficult, and the relationships among the many observation levels which are involved become intricate and largely uncontrolled.

In conclusion, it should be noted that, in addition to the above-mentioned Japanese domes, the history of architecture is full of examples of artificial landscapes. One of the most important examples can certainly be found in 12th century when the doge of Venice, Caprese, had the architect Nero Faggioli (founder of the School of Lattuga where great masters such as Brunelleschi and Ghiberti were educated) build him an artificial mountain landscape, complete with a garden, zoo and even a stream which, driven by a pump, flowed down the mountain.

4 The Artificial as an Interface

While the artificial always needs to interface with nature through *ad hoc* devices, the artificial itself can appear as an interface. This is the case for a whole series of objects or devices, which have been used in the military camouflage technique. Such devices and objects are now coming out on the market especially in the United States. Their purpose is to protect local environments – for example a residence – from the surrounding environment.

The same is true of the products of Larson Utility Camouflage in Tuscon, Arizona. As its entrepreneurial mission, the company declares that it

'(...) wants to help keep utility devices, such as cellular telephone towers and transmission antennae, as inconspicuous as possible, minimizing visual intrusiveness on surrounding neighborhoods and natural environments.

We conceal cellular transmission towers with artificial foliage and bark replicating any tree species, for example lodge pole pines, saguaro cacti and fan palm trees. Larson's architectural facades, crafted of fiberglass or lightweight concrete, can be designed to blend with virtually any building style, while shielding antennae, dishes, masts, poles and towers. Our artificial rock and plants, virtually indistinguishable from nature's own, serve as sound barriers and erosion control for highways.' (LUC, 2001).

Since it is an advertisement, obviously there is no mention of the possible 'adverse side effects' of these products, i. e., the undesired effects or degradation which could occur once they are placed in permanent relationship with the natural environment.

On the contrary, the techniques used in making of 'artificial reefs' are based precisely on these effects. A man-made object is placed underwater – studies take advantage of the observation of sunken ships – and allowed to become a part of the ocean ecosystem. The host environment attacks it with its own forces and aptitudes in order to reduce it to something compatible with itself.

For example, barracudas are known for their ability to extend their own territory to include a sunken ship just moments after it has sunk. One of the main goals of artificial reefs is, as Indiana University researchers report, to draw increasing numbers of swimmers to these structures, in order to lighten man's pressure on the reefs in the natural ecosystem.

Furthermore, since natural reefs are getting rarer in many ocean areas, artificial reefs have also become important from an economic standpoint. They increase opportunities for recreational fishing and swimming sports in coastal waters, as well as the habitat's overall productivity. Steel or concrete structures are placed underwater, as well as pierced fibreglass balloons, which will act as a support for the development of the communities of organisms that normally grow on natural reefs. In theory, and depending on the materials used, artificial reefs could be usable for several centuries. What is interesting here, is that, according to Yehuda Benayahu of Tel Aviv University,

'(...) with time, there is an absolute similarity between natural and artificial habitats' (Kern, 2001).

This is quite consistent with our assumptions: indeed, the natural environment around the artificial habitat – which is intentionally designed without established boundaries – will interact with the device at all the possible observation levels and will eventually transform it according its own rules and requirements. Thus, nature completes, so to say, man's work adding, with time, all the features which were not intentionally designed. On the other hand, this is a very special situation in which the emergence of properties does not come from the artificial as a rationally designed product, but, rather, from its 'degradation' under attack from external natural phenomena which generate just what the designer desires. Something similar happened in a recent experiment on artificial leaves. The study aimed to understand if a particular kind of fungus, which lives on the surface of a particular kind of leaves, could colonize artificial leaves as well.

According to the researchers

'Colonization of artificial leaves also supports previous reports that epiphylls are not species specific in their use of substrate (...), although host specificity may become important under conditions of stress, such as low water availability (...). Microhabitat humidity, light and nutrients transported by rainwater, dust, animals and falling microdebris apparently were enough for successful colonization of the artificial substrate' (Monge-Nájera, Mario Blanco, 2001).

Determining the differences between natural and artificial constructions becomes, in other cases, of utmost importance. For example, barriers needed

'(...) when reprocessing spent nuclear fuel are ultimately to be disposed of in deep underground rock formations. Designing safe and sensible disposal facilities and selecting their sites will require methods by which to accurately assess the performance of deep rock formations (natural barriers) and the glass logs (artificial barriers) encasing the HLW (High-Level Wastes) to be buried ' (Kawanishi, 1995).

Of course, a different 'philosophy' comes into play when man tries to defend a landscape from some natural erosion phenomena by building *ad hoc* barriers. As it has been re-marked

'The public has long considered that artificial barrier dunes are the natural or desired configuration and that the erosion of the shoreline is detrimental. Application of the philosophy of adjusting to and living with the forces of nature will require new efforts to inform the public of the constructive nature of an every changing landscape where erosion of the coast plays a significance role in maintaining the environmental health of these areas ' (Stuska, 1998).

5 The Difficult Choice Between Structure and Process

The exemplar and essential performance always involve an unlimited number of observation levels, but, as we might imagine, designers can only consider the levels they have knowledge of or, among these, the ones that seem more easily approachable both from a scientific and reproductive stand-point.

As we have pointed out, every single artificialization task, whether it be analytical or aesthetic, concrete or abstract, can only take into account one dominant observation level. There are circumstances in which this constraint seems to completely absorb researchers, as in the case of artificial intelligence. But in the field of concrete artificialism as well, the improvement of a design can lead to several subsequent shifts of observation levels, at times giving rise to designs that turn out to be more complex than they first seemed.

The case of artificial blood is typical, in this sense, because its extreme complexity is in no way inferior to other types of organs, tissues or biological processes. Artificial blood must have a suitable fluidity and the ability to provide nutritional supplements like natural blood and, above all – this is the true essential performance of current designs which involve a proper observation level – the ability to transport oxygen. This is the essential performance of current projects. In fact,

'Blood does many things, of course, and artificial blood is designed to do only one of them: carry oxygen and carbon dioxide. No substitutes have yet been invented that can replace the other vital functions of blood: coagulation and immune defense.'

In any case, it is not unusual in the field of the artificial for the reproduced essential performance, to show some improvement over the natural one. As a matter of fact, artificial blood

'(...) will perform their specialized function-delivery of oxygen to tissues-even better than blood' (Winslow, 2001).

The loss of considerable amounts of blood can be compensated by the use of saline fluids which, however, do not possess the main property of blood, the ability to carry oxygen.

At this point, the artificialistic strategies go in two different directions. On the one hand, there may be an attempt to produce compounds that do not carry oxygen directly, but which are nevertheless able to facilitate the solubility of oxygen in blood. In this case, there is focus on the essential performance as such, regardless of the biochemical structure involved in the natural process, namely hemoglobin. On the other hand, research may aim to reproduce this structure directly, and therefore it becomes the real exemplar with its own observation level. Hence, researchers try to understand more thoroughly the finer aspects of the structure in question and how such a structure influences its predisposition towards oxygen.

Naturally, a common problem related to the two tendencies is toxicity. At any rate, the problems associated with the realization of artificial blood are closely related to the problems concerning artificial organs because, in many cases, blood is an inevitable interlocutor. In the case of the artificial heart, at least two physiological systems are particularly sensitive to implantation: the haemostatic (coagulation) and inflammatory systems. Their activation can lead to the formation of thrombi on the artificial device's surface and this, in turn, can cause its failure, an attack on the natural organ or a hemorrhage.

Hence, research is oriented towards the study of thrombotic and inflammatory processes in relation to the materials used in the implantation, while at the same time, it tries to provide treatments of the surfaces involved which minimize these undesired events and suggest improvements for the design of devices.

The retina model, in the case of the artificial eye, seems to be the only alternative, since it is difficult to theorize a device able to perceive images from the outside world which does not have a dot matrix structure. As we have seen in the first part, citing Mahowald and Mead's work, the reproduction of the retina is the main step towards building an artificial eye and it must comply with a few inevitable structural conditions while temporarily neglecting others.

The artificial eye as a structure must, in any case, be combined with the eye as a process, i. e., with what we call 'eyesight'. This process, an essential performance which prevails over all other concerns, is however reproduced under the constraint of the exemplar's structural nature, specifically the retina. In other words, since we must make sure that the retina sends electrical signals that correspond to what the brain – the true center of 'seeing' expects to receive, it must function under the control of an algorithm. The algorithm is carried out by a small computer or suitable pilotage electronic circuit of the retina, which produces the desired signals in the most suitable form.

The retina's essential performance, considered as the exemplar, must then make way for, or adapt itself to, the general essential performance's requirements, i. e., sight as a process. This is therefore a good example of a situation in which artificial very often finds itself: when the exemplar is made up of several subsystems – each of which could be conceived as a separate potential exemplar. In fact, there is always only one overall essential performance which must be fulfilled, and this ends up dominating the characteristics of the various subsystems.

An external artificial retina like the one designed by Mitsubishi Electric (Advanced Technology R&D Center) and by

Mitsubishi Electric Information Technology Center America (MELCO Research Laboratory) called *image sensor*, capable of recognizing a person's head movements and transforming them into input for a computer, will be controlled by an algorithm (*real-time vision algorithm*) which is definitely different from the one that must guide the work of the retina in order to obtain useful signals for the human brain.

In any case, once again, at least ideally, a natural organ, the brain, will be 'deceived' by the artificial device: indeed, the brain will process the perceptive data 'as if' they came from a natural retina. The same thing could be said of the computer connected to the Mitsubishi retina, which will accept the input 'as if' it came from someone typing in commands on a traditional keyboard or from the movements of a 'mouse'.

The Japanese have recently presented the prototype of a machine, also based on the perceptive and processing ability of a variation of the artificial eye, which is able to translate the hand movements of an orchestra conductor into signals capable of modifying various acoustic parameters of an electronic synthesizer (volume, dynamics, color, etc.). This certainly calls to mind – but we will leave it to the reader to unravel the meaning of this analogy – a remark made by Johann Sebastian Bach and related by Köhler:

'(...) the organ? There is nothing special about it: all you have to do is press the right keys at the right moment and the instrument plays by itself (...)' (Köhler, 1776)

6   Artificial Organs and Senses

Canadian Biomech Designs Limited, in cooperation with the University of Alberta and Dendronic Decisions Limited, has recently come up with the design of its C-Leg, an artificial leg for people who have undergone a trauma or the amputation of this limb above the knee. The C-Leg, like many artificial limbs or organs which must fulfill aesthetic and functional requirements at the same time, is the partial analogy of an anthropomorphic robot, in the sense that it must appear aesthetically as a leg and, inside, be able to function as a natural leg. As always, the two observation level levels, aesthetic and functional, are in no way correlated with each other, nor could they be on an analytical level, i. e., the device cannot consider and reproduce all levels, from the epidermis to the osseous cells.

Indeed, the C-Leg privileges, as an essential performance, the leg's dynamic flexibility, i. e., its ability to support the person's body by adapting to the various real situations which he could find himself in. The exemplar, therefore, is the leg as a structure, but its essential performance, intelligent flexibility, dominates the design and the structure is reduced to a physically suitable support, which is light and resistant enough, etc.

The main structure of the C-Leg has nothing to do with a natural leg, since it is equipped with a computer which gives the limb local intelligence whereas, in the natural case, even without excluding the possibility of local control circuits, the organ assigned to such a purpose is the central nervous system. The C-Leg computer controls a valve that stiffens or loosens the knee joint based on a continuous cascade of signals which tell it how much pressure is on the limb. Hence, the presence of obstacles, loss of balance, etc., can be deduced from these signals. In such circumstances, the artificial leg stiffens the knee and places itself in a 'stumble state' in an attempt to avoid a fall. The task of Dendronic Decisions is to design the most suitable software for control of the limb.

While it is easy to grasp the delicacy of such a design, and a whole series of possible side effects (not on a biological level, but on a mechanical one), we must also consider the fact that a person, equipped with this kind of artificial leg, could adapt very well to its operation after a period of training, just like what occurs in the use of many machines which, very often, are operated better by their users than their designers were able to imagine. As a psychological side effect, the person's ability to interpret and predict the artificial leg's based could in fact come into play, adapting to it to such an extent as to compensate for the exclusively local nature of its intelligence. This could start an interesting interactions in which two systems – the man and the artificial leg – attempt to adapt to each other. This sort of interaction was already noted in 1991 by a group of researchers in the field of computer based musical instruments. They considered a rather 'intriguing problem', namely the relationship between the ability of some experimental electronic instruments to adapt to the style of playing of the human being at the same time that the human being tries to adapt himself to the instrument (Lee, et. al, 1991).

The case of the artificial hand is even more complex, if it is to have not only the aesthetic appearance of the various prostheses in existence, but real capabilities of articulation, movement (excluding the use of fingers for the moment). The Utah Arm and Hand System, for example, uses a control (*Sensitive Proportional Control*) which is in turn piloted by the myoelectrical signals of two muscles. This control makes the regulation of elbow, hand and wrist movements possible. It is a device that gives the person with the artificial hand the possibility to move the elbow or wrist at different speed, unlike other prostheses which can be operated solely *on-off* logic, i. e., by activating a switch.

Small stainless electrodes are placed on the skin so as to register the voluntary activity of the two muscles, that is, their myoelectrical signals. These signals are then amplified and used to pilot the prostheses' movement. Everything, as always, is made aesthetically similar to a natural arm and hand.

What differentiates this type of artificial organ from the one seen above is the designers' decision to select, as an essential performance, not so much the elbow and wrist movement – already available in the external electrical drive *on-off* prostheses – but the control of movement itself. The hand, arm and wrist, as in the previous case, only have a minimal structural similarity, apart from aesthetics, whereas the true exemplar is the biological circuit which, in nature, makes control of the limbs as an essential performance possible.

The structure of this circuit, obviously simplified, is rather similar to the exemplar even if the 'degrees of freedom' given to the hand and elbow are subject to of considerable restrictions because the signals, produced intentionally by the person, are nothing compared to the wealth of the natural organs' innervation and musculature which the amputee does not have.

It is clear that many problems are irrelevant when designing organs which will not be implanted in the human organism, but, rather, are destined for machines such as robots that do not have any direct physical connection with man. This is the case of artificial muscles designed by American SRI International, based on the use of films of electrorestrictive polymers (which contract or stretch under electrical tension) to be used by a small robot. Artificial muscles, according to SRI reports, are comparable to natural ones as far as performance is concerned, but are more efficient and, above all, faster.

Also research on artificial muscles also includes the possible transfiguration of the natural exemplar, in this case an amplyfying transfiguration, because in BAM project

'Shahinpoor's research documents state that the fibres are capable of holding four kilograms per square centimeter. A human biceps can lift a maximum of just over two kilograms per square centimeter' (Gibbs, 1998)

The natural exemplar's structure, the culmination of millions of years of evolution, is never overlooked by designers who may, however, give up its reproduction when it is too complex, unknown or made of materials which are impossible to reproduce with existing conventional technology. In these circumstances, researchers' efforts are directed to reproducing the exemplar's structure only in analogical terms. In a way, the strategy of nature is accepted, but its operating principles are modified to a varying extent in order to obtain the desired essential performance.

This is what is happening in the design of the artificial nose or, rather, of artificial smell. The natural nose possesses millions of receptors, made up of proteins placed on the surface of the cells which are able to capture the molecules of odors. Nevertheless, there are only 10,000 types of receptors, a number which is not sufficient to recognize all the odors that we are exposed to. The brain recognizes odors by examining the combinations of receptors that each particular odor activates. Since there are millions of possible combinations, the brain is virtually capable of distinguishing the odor of all existing molecules.

Researchers' attempts to reconstruct the sensible 'repertory' of the natural nose had not gotten very far until very recently when John Kauer and David Walt from Tufts University found a completely new approach. It consists in substituting the natural strategy based on chemical recognition with a strategy based on optical recognition. The artificial nose is made up of a system of 19 optical fibres covered with different fluorescent compounds containing a dye called *Nile Red*. When the molecules of a given smell hit the fibre cover, it emits a fluorescent light with a different intensity and wavelength from the one emitted by the nearby fibre and, therefore, every possible odor produces its own fluorescent spectrum.

At this point, the spectrum is transducted into electrical signals and these are coded so they can be interpreted by a digital neural network, i. e., a software system capable of recognizing the various configurations that the data can have, by classifying them into sufficiently homogenous 'types'.

This is an artificial device which, perhaps more clearly than others, reproposes the criterion of functional equivalence which we mentioned in the chapter dedicated to artificial classification. It would obviously be absurd to study the artificial nose in order to gain a better understanding of human smell, since the receptors' operating principle is completely different and the analysis of the information from the combinations of receptors is carried out on an informational basis (by the computer) and not on a biochemical basis as occurs in the brain. Nevertheless, the debate remains open regarding the possibility that, at least, the 'logic' of the natural brain might be reproduced by explicit algorithms or by the self-organization of digital neural networks.

At any rate, the transfiguration of the exemplar's structure and the ways in which it carries out its essential performance allow the artificial nose to carry out tasks which are completely impossible for the human nose, such as monitoring chemical changes inside the human circulatory system.

The fact is that the need for more and more refined artificial sensors in the field of robotics makes the observation and knowledge (and therefore the modeling) of natural sensors ever more important. So, among others, even a European research group coordinated by the DIP-INFM of the University of Genoa is developing a research project on natural and artificial sensors in order to enhance our knowledge of the first stages of information processing in visual, olfactory and auditory processes to improve the devices which robots must use.

The natural processes assumed as essential performances are phototransduction (for sight), chemiotransduction (for smell) and mechanical transduction (for hearing) in vertebrates, whereas the applicative objective is the design of autonomous 'navigation' strategies for mobile robots.

Significantly, the research also intends to

'(...) compare natural sensors and artificial ones in order to establish which characteristics of biological sensory processing can be useful for the designing of new artificial sensors or robots' (DIP-INFM, 1995), an aim which is at the center of bionic programs. In the various cases which the project deals with, reference exemplars include, the fly and several amphibians.

The artificial heart, has also been the subject of several different reproduction strategies. The structure of the exemplar and its essential performance (the pumping of blood in the organism's circulatory system) carry out different roles in the various projects. At Kolff Laboratories in Salt Lake City, for example, and in many other research and designing places, the anatomical structure of the natural heart constitutes an important aspect, because the reproduction of the essential performance associated with it depends considerably on the reproduction, even if it is modified, of the various cardiac subsystems: from atria to valves.

On the contrary, privileging the essential performance provides a new example of innovation regarding to the exemplar's structure and also a specific aspect of the essential performance itself, namely the pulsating activity of the pumping of blood. This is the case for the *Streamliner* project of the McGowan Center at the University of Pittsburgh, which is expected to be realizable within just a few years. It is a device which, when completed,

'(...) will permanently substitute the majority of the natural heart's pumping functions' (McGowan, 1999)

The innovation consists in the fact that, instead of reproducing the usual essential performance by technical contracting devices capable of generating a rhythmic and therefore pulsating pumping, *Streamliner* is designed to pump continuously and produce a continuous flow thanks to an electrical engine which rotates at 10,000 revolutions per minute. Hence, the recipient of such an artificial heart will not have pulsations.

As researchers at the McGowan Center maintain,

'(...) although *Streamliner* is different from a natural heart, it will take advantage of the sturdiness – and absence of various weaknesses – of the biomaterials it will be made up of. For example, thanks to the way in which an electrical engine works, it is much easier to design a constant flow compact device rather than a *bellows type* which requires compressed air lines that can allow infective biological entities to enter the organism' (ibidem).

There is considerable interest in this project, because the transfiguration of the exemplar, but above all its essential performance, will likely produce various side effects which are now unforeseeable, apart from a few psychological effects, which we can already imagine, given the absence of pulsations.

Regarding transfigurations or side effects deriving from the materials and procedures adopted in the design, it is important to observe how the artificial kidney – another great protagonist of bioengineering – is able to remove both bromide and iodine just like the natural kidney. But the artificial kidney can also distinguish them, and thus remove them according to their gradient. Moreover, the artificial kidney is able to remove potassium in a very short period of time.

The complexity of the exemplar's structure and its performances is daunting at times, compared to our current knowledge and technological capabilities. Apart from the brain, the liver seems to be the most complex organ to approach and reproduce for the moment. It is a kind of chemical system that functions 24 hours a day and is responsible for the production, storage, metabolism and distribution of a considerable amount of basic nutrients for the organism's health. Besides producing glucose and proteins and factors which help the blood to coagulate, the liver acts on waste products and transforms some of them into usable elements, while discharging the ones that are dangerous. At the same time, the liver produces vitamin A and stores vitamins A, D and B12. In short, a chemical firm would need a plant stretching many hectares to do a similar job.

Clearly, before such a wide spectrum of performances, the selection of one of them as essential is prohibitive yet the reproduction of all of them would require a model, to design and coordinate the interfaces, which is presently beyond our capabilities. This explains why it is better to resort to what we could call a mixed strategy for this organ: this is the case of the missnamed 'artificial liver' under study at Edinburgh Royal Infirmary and the University of Glasgow Strathclyde, which uses pig cells. A similar project, in the development phase at Queen Elizabeth Hospital in Birmingham, uses residues of human liver cells derived from transplant operations.

Interest in the latter project stems from the fact that the plasma, sent to the device through a tube inserted in the groin, passes through a network of tiny permeable plastic tubes which force it to mix with the cells, thus removing toxins and converting the necessary compounds. A second network of tiny tubes then restores the treated plasma to the organism.

The composite nature of this object makes it a typical example of hybrid artificial, in which a direct attempt is made to match a technological subsystem, the network of tubes, with a completely natural one, namely the liver cells. The essential performance can proceed undisturbed by reactive phenomena in the organism because it is realized on the outside and with a structure of the exemplar which is free from isomorphic constraints.

A very similar hybrid strategy, which, however, involves placing the artificial device inside the organism, and therefore compliance with some structural analogies with the exemplar, is used in the development of cardiac valves. In particular, for the valves designed by MASA (Mid Atlantic Surgical Associates, New Jersey) researchers use tissues taken from ox or pig pericardium while the supporting structure is realized by polyester plastic materials. The advantage of using animal valves stems from the fact that, in such a way, one can avoid using anticoagulants to compensate for the organism's reactions against foreign materials.

Moreover, the same researchers, adopt a completely artificial strategy in other cases, above all when the patient is younger and can bear the use of anticoagulants (or when the patient already uses them for other reasons). For this type of valve the same ceramic material (which lasts a very long time) used for the covering of the space shuttle is employed; however it requires the use of anticoagulants.

Finally, there is a long series of products which are midway between bioengineering and a technology aimed above all at restoring the body aesthetically – which we will deal with in a later chapter. These products include hand, wrist, shoulder and foot joints. Normally, at least in devices of greater bioengineering interest, silicone elastomers are used for their high endurance and their good biocompatibility, as well as titanium for the construction of the supporting parts.

7   The Artificial Brain

Although the expression 'artificial brain', which was somewhat accepted in the 50's and 60's, is now obsolete, artificial intelligence research seems to be once again heading towards this objective after more than thirty years of attempts in which terms such as 'mind', 'reasoning', 'understanding' and 'intelligence' have been used instead of 'artificial brain'.

By this we do not mean to assert that it is now time to go from pure metaphor to the concrete realization of an artificial brain. What we mean, on the contrary, is that research on artificial intelligence has led to the growing belief that the physical dimension – meaning all of the not merely informational conditions that characterize the work of the mind – not only cannot be overlooked, but is probably one in the same with the performance of intelligent activities in man. In short, even the

most abstract intelligence, is not born out of physical nothing-
ness but is produced in relation to the continuous exchange of
matter and energy with the organism's internal and external
environment. In other words – and putting aside metaphysical
considerations without denying their importance – the brain
appears more and more to be what it actually is: an organ like
any other organ, but, considered in its whole, much more
complex.

Over the past thirty years, artificial intelligence has been
called 'symbolic', since it maintained that it is possible to reproduce intelligence as an essential performance of the mind,
based on manipulations of symbols while completely overlook-
ing the structure and physical components which are nevertheless very obvious and important in nature. In fact, researchers trying to understand brain functioning, are taking
many different paths. For instance, at the University of Southern California biomedical engineers try

'to treat the whole system (the brain) as what engineers call a
"black box". In addition to trying to understand what goes on inside
the box, we have tried to build a box of our own that behaves in exactly the same way, an electronic device that responds to any given
input signal with precisely the same output signal that the natural
neural system does' (Berger, 1995)

Such an approach did not have any metaphysical understanding: artificial intelligence researchers do not think of the
mind at all as metaphysicians would think of the soul, nor do
they consider intelligence as one of its qualities. Rather, with
an orientation which we can call functionalistic, they thought,
and think, of the mind as the brain's software. Since software,
as an informational structure, is transferable *ad libitum* on
other physical supports, it can be inferred that even thought,
understanding, reasoning and intelligence can be reproduced
at will on any other supporting structure, for example the
computer.

Therefore, the foundation of this approach, lies in the above-mentioned principle of functional equivalence. Authors such as P. N. Johnson-Laird and Z. Pylyshyn have thoroughly explored this subject, the true cornerstone of the whole artificial intelligence case or, at least, what J. Searle defined as 'strong' artificial intelligence (Searle, 1984), according to which thought ability actually takes place in the computer.

In order to further clarify the principle in question, it is worth reading carefully the following quotation from the physicist Craik, quoted by Johnson-Laird in 1988:

'When we speak of model we thus mean every physical or chemical system whose relationships are structured as in the process which is imitated, that is, in which there is an equivalent structure of relationships. By "structure of relationships" I do not mean a mysterious non-physical entity which accompanies the model, but the fact that there is an operating physical model which functions in the same way as the process which is parallel to it'. (Johnson-Laird, 1988)

Now, if we consider the actual results and products which artificial intelligence has produced throughout these years, we realize how groundless and useless for the discipline's progress those claims were.

The most efficient computer programs, inspired by artificial intelligence models, are certainly expert systems, software capable of reproducing human reasoning – of an expert in a specific field: medicine, law, geology, etc. – for example, of a diagnostic type, by operating on two elements: a 'knowledge base' and an 'inferential engine'. Whereas in the former the knowledge of human experts is coded, which makes it available through meticulous and complex interviews, the latter consists of an algorithm capable of performing deductions, inductions and probability calculi.

In order to illustrate these concepts let us consider a general example. If the concept that 'illness M is identified by symptoms A, B, C, D,' is inputed into the knowledge base, the expert system, which is required to provide a diagnosis for someone with symptoms B and D, will ask if symptoms A and C are also present and, if this is the case, it will assume illness M. The expert system could present the same diagnosis even without symptom C, if the human expert who instructed the computer gave it less or additional importance compared to A, B and D.

Clearly, such a system can be presented in much more complex terms according to the amount of knowledge that is at the expert system's disposal and the various methods of analysis and prediction it is equipped with.

Nevertheless, a few objections can be made that, of course, do not concern the usefulness of these programs, but rather the pretence that an expert system reproduces the mental activity of a human expert. Deductive activity, based on an appropriate representation of knowledge, is certainly a performance of the human mind, but why should it be the 'essential' one? The fact is that, as M. Minsky and R. Schank – two artificial intelligence pioneers – have observed on many occasions, every model, in working on human intelligence, offers something useful but, nevertheless, something which escapes its attention, sooner or later, determines its limits and inadequacy.

Put in our lexicon, we must acknowledge that, when designing a model of the mind or even only of intelligence, every observation level offers some advantage, to the detriment of other aspects which invariably prove to be crucial on various occasions. Thus, as regards intelligence, the limits of expert systems lie in their inability to receive, and thus reproduce, intuitive, synthetic or very personal knowledge, which however constitutes a good share of the intelligence that a human expert resorts to. This difficulty cannot be attributed to a particular defect in programming the machine: rather it is due to the fact that the human experts themselves are not always able to explicitly state the line of reasoning they employed in developing their diagnosis, prediction, etc. In these circumstances we speak of 'tacit knowledge' (Polanyi, 1966), which includes all the mental abilities that, although they are real and produce real effects, cannot be completely expressed even by the person who possesses them. Other strategies, such as learning by example – which, to a certain extent, should reproduce the learning that takes place when a good assistant works next to an expert – have not improved the situation very much.

This sort of limitation can be attributed to the fact that computers are supports and processors of information, and not knowledge. Production of knowledge implies information processing but information processing, as such, does not imply *ipso facto* a production of knowledge. More importantly, the rules according to which we process information – which can easily be transferred to the machine – have nothing in common with the ways we think. Besides, the informational observation level *is* important in the mental world, but it is not easily isolatable from the whole of events that characterize it.

After all, even a simple electronic calculator processes data by developing mathematical calculi, and the deduction is nothing but a calculation. Once it is isolated, let us say almost eradicated, from all the dimensions of mental life – which are moreover unknown in their entirety – the ability to develop mathematical or logical calculi is presented as a very relevant but arbitrary essential performance: it is very useful where the problems to be solved are compatible with it and completely insignificant where the problems are of a different nature, such as those related to several forms of common sense, intuition, criticism, suspicion, etc.

Since artificial intelligence has never been able to seriously consider the idea of reproducing the structure of the natural exemplar (the brain), it is clear that its falling back on the principle of functional equivalence refers to some essential performances in themselves. In this sense, artificial intelligence has been successful, just as pure and simple automatic calculus has been successful. The different hardware and software support, – or, in our lexicon, the different materials and procedures – adopted to realize intelligent devices has nothing to do with what Craik calls 'model', in which

'(...) the relationships are structured in the same way as the process which is imitated (...)' (Johnson-Laird, cit.),

Such is the case for the simple reason that the mind is a largely unexplored model and the object, moreover, of social representations which are strongly influenced by current doctrines, extra-scientific paradigms, etc.

On the contrary, it is a plausible idea that, as in many other cases of artificial objects, processes or machines, even in artificial intelligence's best results we note the definite priority given to essential performance. Even if it is only of an informational nature, it is an aim to achieve, and in fact pursued in various aspects. Indeed, it would be difficult to deny that an expert system is intelligent, at least within the limits of intelligence understood as the ability to use information rigorously. Hence, we can claim that artificial intelligence produces real intelligence but also 'authentically artificial' intelligence.

Towards the end of the 80's, in the wake of the disappointment regarding the lack of success of 'strong' artificial intelligence, a research tradition which dates back to the 50's – but which has its foundations in the logical works of McCulloch and Pitts in 1943 – came back to the fore. This is the tradition of the so-called artificial neural networks. Even in this case, reference to the exemplar – the brain as a structure of physical connections between neurons – is little more than evocative, even though research on neural networks, unlike symbolic artificial intelligence, can rely on some scientifically established knowledge. Basically, as we have said before, the neural networks reproduce the model we have of the connections between the neurons which characterize the brain, but, obviously, instead of the structures, tissues and biochemical signals of this organ, the components adopted are electronic information units and mere electrical signals.

According to the general definition advanced in 1988 by DARPA (Defense Advanced Research Projects Agency), a neural network

'(...) is a system composed of many simple processing elements operating in parallel whose function is determined by network structure, connection strengths, and the processing performed at computing elements or nodes' (AFCEA, 1988)

According to Haykin,

'A neural network is a massively parallel distributed processor that has a natural propensity for storing experiential knowledge and making it available for use. It resembles the brain in two respects:
1. Knowledge is acquired by the network through a learning process.
2. Interneuron connection strengths known as synaptic weights are used to store the knowledge' (Haykin, 1994)

Since even in the human mind learning seems to imply strengthening the synaptic relationships between neurons, we can say that the neural networks reproduce the exemplar's structure, namely the architecture of the brain, at least at this observation level.

The usefulness of neural networks lies in their capability to solve problems for which there are no symbolic algorithms or these algorithms are too complex to design. As paradoxical as it may seem, it is actually true that the symbolic or mathematical description of the process through which we quickly recognize our car in a crowded parking lot without any calculations is much more complex than the description and calculation of an astronomical event. Neural networks thus try to simulate the identification process by imitating rather the brain's automatism – or better yet its self-organization – rather than an approach which calls for a formal and conscious analysis.

The idea is that the presentation of a certain data configuration to the network – the numbers which describe a certain computerized picture, for example – strengthens certain connections between neurons rather than others and therefore similar configurations strengthen the same region of connections more and more. Hence, an appropriately trained neural network is able to distinguish a specific configuration (for example a fingerprint) from others or from the background. It should be noted that the network is governed by mathematical criteria of calculus (of the relationships), but it does not follow any algorithm, that is, it does not perform any program, as does occur in symbolic artificial intelligence.

It is evident that these two currents of thought on which artificial intelligence is based, privilege two groups of essential performance which are different from one another, yet present in our brain at the same time. Furthermore, it does not seem at all easy, as we have maintained with regard to all artificial objects, to find a way to get a neural network and a symbolic artificial intelligence software to work together, without having to face the problem of finding and describing a superior model and, possibly, the natural exemplar it refers to. The latter should include the first two, not in an arbitrary manner, but in a way which is as similar as possible to what happens in the natural brain, in which performances of recognition and formal, informational or even symbolic performances are certainly present and coordinated amongst themselves.

A recent research development in artificial intelligence foresees the opening of this discipline to the social sciences. It is a new approach which, considering the fact that human intelligence is developed in the environment, regards the social relationships between software or hardware 'intelligent agents', as models of study and design. Some of the main topics regarding this approach, to research on components of social relationships, include: credibility, imitation, cultural adjustment and the co-ordination between the internal or external dynamics of the agent.

The model is therefore made up of society, in a simplified or implicit version. However, the essential performance varies from project to project, leaving open, as usual, the problem of their co-ordination in an integrated framework. Such a framework, could only be found in some general theory of social relationships, namely in a field in which sociologists themselves have been working for at least a century without accomplishing anything definitive.

In any case, man's role itself becomes odd at this point, both as a designer and as a user, since, at the moment of designing and using similar intelligent agents, he will inevitably become related to these agents, setting off unpredictable phenomena of mutual adjustment. This is a topic which concerns many types of intelligent artificial devices, in which man himself becomes a part of the process he designs (*human in the loop*) by trying to give the machine the ability to adapt to human based.

In conclusion, all the cases of artificial intelligence which we have examined, are evidence, on one hand, of the ambition and necessity to find useful models in nature – in man, in society – in order to design intelligent machines or processes and, on the other hand, the inevitable transfiguration of those models and their performances by the artificial.

For example, an expert system not only reproduces some of a human expert's explicit abilities, but it can do so with extreme speed, accuracy and without ever forgetting those abilities, even in environmental situations in which man would find himself in serious difficulty. Furthermore, an expert system can draw on a wider range of expertise, making use of the skills of different human experts. Such a system can therefore operate as if it had more than one personality, precisely because it has no personality and its way of reasoning, unlike man's, can be varied *ad libitum*.

Neural networks, but the issue also concerns the 'sociological' perspective of artificial intelligence on a different level – often lead to completely new diagnostic or recognition behaviors, whether they be erroneous or simply 'mysterious'. Hence, they offer the possibility of speculating on alternative ways of interpreting and interacting with reality, on the basis of which new models and styles are already being developed in the field of play, art and the design itself of machines destined for various non standard functions.

Likewise, ALife programs, although they were born out of the desire to capture the essence of natural life and evolution through algorithms, end up providing researchers with models of life '(...) as it could be' (to use a famous expression by Chris Langton) – as well as models of life for entertainment products and other kinds of products. In conclusion, in this field of the artificial as well, even though the objective of the reproduction of the exemplars and their essential performance is only partially achieved, 'something happens' and it should not taken for granted that this something is less interesting or important than the unsuccessful reproduction objective.

8   Prostheses and Surrogates

The category of artificial objects, processes and machines of a purely aesthetic kind – therefore without any pretence of reproducing the exemplar as far as structure or functionality are concerned – is as ancient as it is vast. It includes fields that range from substitutive prostheses of organs or body parts to the reproduction of natural components of landscape, from the reproduction of nature's scents to imitations or 'false' products, for commercial purposes, which get their inspiration from pre-existing technological products. For instance, as far as leather is concerned, it

'(...) was first created in the U.S. in 1963 and was later intro-duced to Japan, marking dynamic growth. Up until the 1970s, artificial leather was considered an alternative to real leather, and in the 1980s it was considered healthier and cleaner than the real thing. Then in the 1990s, when the necessity of conducting environmental countermeasures on a global scale became apparent, artificial leather manufacturers strove to develop products that would place less strain on the environment' (Teijin, 2000).

Although there is no room in this work to discuss it at length (a more complete examination can be found in the bibliography at the end of the book), it is useful to keep in mind that art itself – just think of figurative painting and sculpture – has a clear ancestral reproductive role. Indeed, we cannot deny that all artists intend to reproduce objects and processes from the outside world or their own mental states as natural facts, interpreting them according to their own poetics, using materials and procedures which are different from the ones that we find in an event in nature or the mind.

The only basic difference between the reproduction of a figure in a painting and a robot in a technological field stems from the markedly different views of the artificial compared to the natural regarding transfiguration.

While for a robot designer, the inevitable transfigurations of his product with regard to the exemplar and its perform-ances are often a source of disappointment and frustration or even of astonishment over something unexpected, the main aim of the artist's work is precisely that transfiguration of the exemplar and its performances. Indeed, even the most 'realis-tic' artist reproduces the world with a deliberate interpretive intention and not as a mechanical copier. Moreover this sort of mechanical copying is impossible to imagine for the reasons which apply to all artificialists. The beauty of a painting lies in the persuasiveness, power and elegance of reality and not its reproduction of reality as such.

The aesthetic level on which reconstructions and various types of gadgets are found is naturally different and, here, the designer's ideal ambition is to reproduce reality precisely 'as it is', that is, as it is perceived and represented collectively. As we know, various parts or organs of the human body have always been the subject of these kinds of reproduction attempts, not only for personal reasons, but also because in many civilizations and historical periods mutilation was feared even more than death, and it was often full of negative meanings.

It is said that Rig-Veda, an ancient Indian poem, contains the first reference to a prosthesis. Written in Sanskrit between 3500 and 1800 B.C., it tells the story of a warrior, Vishpla, who, having lost a leg in battle, was fitted with a metal leg so that he could continue to fight. Likewise, the Celtic god New Hah was said to have four silver fingers.

Wooden, bronze and leather prostheses have also been found in Roman ruins dating back to 300 B.C. and Pliny the Elder, in the first century B.C., tells about a Roman general, Marco Sergio who, after the amputation of his hand, of his hand, had a metal hand built to hold his shield and continue his military activity. Subsequently, in the Middle Ages, all kinds of prostheses made their appearance, both for military purposes and in order to hide a deformity, and the Renaissance, giving new impetus to the scientific study of nature, renewed the efforts which had begun in ancient times.

Nevertheless, as a rule, this area of artificial objects now has a very particular relationship with bioengineers' concrete-analytical reproduction attempts which is also almost always oriented towards the maximization of the aesthetic effect.

The human eye is perhaps one of the most interesting examples. The traditional prostheses which we know about – invented in 1938 by an American doctor, Fritz W. Jardon – are often designed very well, but limited since the eye-ball is motionless while the normal eye-ball is in constant movement.

But there is now a new kind of artificial eye, designed by Bio-Vascular, Inc. Thanks to a special biomaterial, hydroxyapatite, muscle tissues grow on the back of the prosthesis and these tissues, with the aid of additional mechanical devices, allow the artificial eye to move simultaneously with the natural eye. In short, it is a case of double illusion. The organism, moves the artificial eye-ball with its muscles 'thinking' that it is the natural one while at the same time, an outside observer is led to believe that the person with the prosthesis has completely normal eyes.

Obviously, this kind of artificial eye does not become an actual part of the organism and various precautions and maintenance activities are necessary. At any rate, the majority of these prostheses last 8-10 years because the form of the eye socket changes over time hence the device no longer fits properly.

Today, hair, above all, constitute an artificialization objective which is pursued very actively. In light of the immunity problems associated with natural hair transplants, artificial hair technology has developed extensively. Above all, it makes use of synthetic materials such as nylon, PET (a polyethylene) and modified acrylic substances. The goal is to reproduce the model in its exterior appearance – perhaps the essential performance selected almost the time – but there is also another consideration, namely the flexibility of natural hair.

As is so often the case, the main problem stems from the relationship with the human body. Considerable efforts are being made here as well, including the use of silver to cover the hairs, thus reducing the chance of infection. Nevertheless, all artificial hair is eventually rejected as an extraneous body. In light of such an ever-present and intrinsic danger, in 1984 the American FDA declared the implant of such devices illegal, while it is still legal in Japan, Mexico and Europe, though surrounded by numerous medical controversies.

Proceedings to artificial objects which do not assume human body parts as models but rather natural objects or processes, in the food sector we find considerable efforts to 'surrogate' natural substances or products. Among the thousands of examples which we all know, just think of the so-called sweeteners or artificial sugars, whose growth was given further impetus after reports were issued on the negative effects of natural sugar. Although such reports were rather exaggerated giving rise to one of the many collective missions full of hysteria, the production of artificial sweeteners has generated various products, none of which, however, has come on the scene without producing unpleasant or harmful effects owing to the principle of inheritance. Saccharine, for example, gives a pleasantly sweet taste, but then leaves a bad aftertaste on the palate; aspartame, in turn, does not have a bad aftertaste, but it seems that aspartame/phenylalanine's metabolites (one of its components) include several toxins, such as methyl alcohol, which can be dangerous even in small quantities.

However, some people, such as Californian Wholesale Nutrition, maintain that artificial honey, which is made from ascorbic acid and various essences, is cleaner and less harmful than natural honey, since that it does not contain adulterants, botulin spores, drugs, pesticides, allergenic factors, bee residue or waste, or other non-soluble elements that come from beehives. In short, as in giving the recipe for the product, a Wholesale Nutrition expert maintains

'(...) we offer our kit to bring our clients' and the world's attention to the insidious nature of all sugars. Without our kit, nearly all of us would continue to ingest about 45 kilos of saccharose, glucose and fructose a year with great harm for our health' (Wholesale Nutrition, 1996).

But efforts to cope with nature cover the whole spectrum of flavor, and, as said by a manufacturer,

'Through careful analysis of key flavor ingredients, Flavor Concepts can help develop natural and artificial food flavorings rivaled only by Mother Nature herself. – or we can create new flavors, combinations and tastes she never even tried' (Flavor Concepts, 1998).

Differences between the taste of natural and artificial products are sometimes intentionally reviewed by newspapers, as is the case for the beer. According the San Jose Mercury-News, artificial beer

'(...) It has a refreshing taste, though a bit sweet, and is best when mixed with extremely cold water. Its taste is remarkably similar to beers produced by micro-breweries. It is a quick source of liquid carbohydrates, and it is easy and light to pack and mix; (...) In very cold water the mix clumps up unless you add water slowly and stir constantly; Although it does not compare to a fine lager, it suffices quite nicely when your taste buds crave a cold one in the backcountry and you do not fancy carrying a six-pack. The manufacturer mentions one can add clear grain alcohol or vodka to achieve an alcoholic beer' (SAN JOSE MERCURY-NEWS, 1997).

In other special circumstances where natural phenomena are well understood and easily reproducible, man is able to get very close to the exemplar and its essential performances. Diamonds are a case in point:

'Most of pure diamond's fundamental properties are retained in artificial diamonds. For example, artificial diamonds have extreme hardness, broad transparency, high thermal conductivity and high electrical resistivity' (Weintraub, 1998).

Likewise, various movements and manufacturers, say 'no to artificial tastes', 'no to artificial colors', 'no to artificial sweeteners', 'no to artificial preservatives', and fight their battle for a return to nature. Whoever has any doubts about the distinction between natural and artificial should find a lot of material to reflect on in this almost ideological open campaign.

Although all the products we ingest are actually made from elements, compounds and substances that exist in nature, the distinction between natural and artificial is apparent to everyone. Such a distinction is easy to make since the combination – good or bad – which nature gives to the elements is by definition different from the one man creates when he attempts to reproduce it as such and, obviously, when he actually attempts to improve it.

9   Artificial Environments

Among the objects of nature which are today most often assumed as models, we can find elements of the landscape or climate, such as rock, snow, grass, rain, islands, caves, mountains, ponds, lakes and many others.

Artificial rain – an undertaking which assumes a process, rather than a structure, as its exemplar, just like artificial insemination – has a rather long history, at least in Kansas where in 1890 a certain Melbourne, 'The rain wizard', shot mysterious gases from his roof in order to stimulate the meteorological phenomenon in question. As far as we know, Melbourne, did not possess great secrets and his practices were not much different, apart from references to something 'scientific', from the magical practices of rain propitiators. However, in spite of his failures, a few private artificial rain companies came into being, each one claiming to have obtained from Melbourne the technology he used.

In Kansas, the Western Kansas Weather Modification Program has been in operation since 1975, and one of its goals, besides artificial procedures to prevent the formation of hail, is to increase annual rainfall. There is even a local museum on this topic where one can examine various apparatus, including a device for the insemination of clouds. The rules of 'cloud seeding,' were discovered in Schenectady, New York, by lrving Langmuir, Nobel laureate in chemistry in 1932 and director of General Electric's research laboratories 1946.

The rainbow is one of the most spectacular natural phenomena humans have tried to simulate. Today, several techniques are available, ranging from very simple means you might find in a physics classroom to very sophisticated technologies. Among the latter, the Hitachi Diffraction Gratings at the Okazaki National Research Institute, National Institute of Basic Biology. According to Hitachi

'The spectrograph has successfully realized the world's largest artificial rainbow whose intensity is 20 times of the sunlight energy right above the equator! Hitachi plane diffraction gratings consisting of varied space grooves have also been adopted in the spectrometer of the extreme ultraviolet explorer launched by NASA' (Hitachi, 2001).

Unlike rain and rainbows, artificial snow has a pragmatic purpose at times and an aesthetic purpose at other times. As we know, the main pragmatic function of artificial snow is to make the sport of skiing possible. The undertaking adopts the structure of snow as an exemplar. This snow is produced according to different principles than natural snow, using water atomizers and compressors. The result is almost always satisfactory, although, according to some northern European skiers, artificial snow produced in the southern parts of the continent, unlike the one available in northern European countries, often tends to be 'pure ice which is used out of despair'. In other cases, focusing more on the functional equivalence principle, the sporting aim is pursued by choosing a more restricted essential performance – the viscosity necessary for skiing – regardless of the production of snow: this is the case for artificial ski slopes made using materials such as polyethylene, which lasts longer and guarantees an ideal flowability with limited friction.

Conversely, the aesthetic purpose calls for a type of artificial snow which would be difficult to ski on, since the selected

essential performance only concerns its appearance. Once
again, the same model lends itself to two different versions of
artificialization, without any possible relationship between
them. In the aesthetic case, it is a fibrous product, readily
available on the market, which can be deposited on the ground
to simulate the presence of snow and, once deposited, as the
producer of one of these substances affirms, you can create
various effects, from footprints to tire prints. However, there
are considerable side effects lying in ambush, in particular,
the danger that the fibres can drift in the air and cause irrita-
tion if they are inhaled.

Another field worth mentioning, which assumes a process
rather than an object as its exemplar, is so-called 'artificial
weathering'. It consists, for example, in monitoring certain organic coatings of various kinds of natural objects by subjecting
them to regular and accelerated conditions using ultraviolet
rays. Normally, as the Swiss Federal Laboratories for Materials
Testing and Research attest, there are not any great similarities between weathering obtained by artificial methodologies
weathering caused by nature, due to the numerous variables
included and the presence of very complicated decomposition
processes.

We should also mention the field of 'special effects' in
films. This field, called fiction, has always made ample use of
devices which tend to give the spectator the most realistic im-
pression of all kinds of situations and events. Specialized
businesses, dedicated to meteorological events, like the ones
produced by American Sturm's Special Effects International,
are able to reproduce winter the whole year long, regardless of
atmospheric conditions or climate. Moreover, from the lightest
shower to the strongest storm, from the lightest wind to cy-
clones, specialists can simulate the climatic conditions of any
season. Of course, in order to appreciate these reproductions,
one has to place himself at a very special observation level, in

order to avoid 'breaking the rules' which allow the reproduction to appear as a natural phenomenon.

Another ambivalent analysis can be made on artificial grass, which is useful in some sports but also made for decorative purposes. The Ten Cate Nicolon product, called Thiolon Grass, uses particularly steady polymers with a high thermal conductivity and therefore a great ability to absorb heat. Hence that, according to the producer, artificial grass made in this way has characteristics which are actually better than those present in natural grass. Indeed, it can be used regardless of the temperature of the place where it is installed; it is produced in several varieties according to the sport; it cushions falls better than natural grass; it is resistant to ultraviolet rays and, of course, it does not require artificial fertilizers or herbicides. No undesirable effects are mentioned, though the producer says he is sensitive to environmental problems and he makes sure that the pigments used are well tolerated by the environment.

With regard to artificial herbicides or fertilizers, there is also the long-standing controversy, concerning possible health risks associated with their use. In principle, we cannot claim that something is harmful only because it is artificial. Whether natural or artificial objects or processes are harmful or advantageous is determined by their structure, that is, by the above-mentioned combinations and recombinations of the same natural elements and not because they are the work of nature or man. The ancient fear of the 'devilries' carried out by technicians – or scientists – today is still a difficult enemy to defeat. Overcoming this fear, however, can at times leave too much room for enthusiasm for a reproduction or improvement of the natural world, by means of technology, without any *a priori* guarantees.

Natural landscapes, in turn, have been altered by man for purposes which often have a prevailing conventional technological meaning – a city, a road, a bridge – but which often try to locally reproduce natural configurations assumed as exemplars for economic, tourist or military purposes. This is the case of the crannogs, or artificial islands, built in Scotland probably during the Neolithic age and inhabited up until the XVI century. Several English university studies are now trying to clarify how the geological and environmental conditions influenced the construction of these islands and what relationships their inhabitants had with their natural surroundings. Likewise, in Japan, the Kasai artificial beaches were built to the extreme north of the Bay of Tokyo and, here as well, research is under way to establish what chemical and dynamic changes are affecting the bottom of the sea following this human intervention.

On a purely aesthetic level, but with all the social and cultural functions associated with it, the technology, or art, of gardening, deserves a separate comment. This technology, which since ancient times has created real local artificial landscapes, involves relationships with the natural context that are often essential.

With regard to gardening, just consider a passage by Edgar Allan Poe from his *The landscape garden*, written in 1850, in which his interlocutor, Ellison, quotes a writer who expresses himself in this way

'There are, properly,' he writes, 'but two styles of landscape-gardening, the natural and the artificial. One seeks to recall the original beauty of the country, by adapting its means to the surrounding scenery; cultivating trees in harmony with the hills or plain of the neighboring land; detecting and bringing into practice those nice relations of size, proportion and color which, hid from the common observer, are revealed everywhere to the experienced student of nature. The result of the natural style of gardening, is seen rather in the absence of all defects and incongruities – in the prevalence of a beautiful harmony and order, than in the creation of any special wonders or miracles. The artificial style has as many varieties as there are different tastes to gratify. It has a certain general relation to the various styles of building. There are the stately avenues and retirements of Versailles; Italian terraces; and a various mixed old Eng-

lish style, which bears some relation to the domestic Gothic or English Elizabethan architecture. Whatever may be said against the
abuses of the artificial landscape-gardening, a mixture of pure art in
a garden scene, adds to it a great beauty. This is partly pleasing to
the eye, by the show of order and design, and partly moral. A terrace,
with an old moss-covered balustrade, calls up at once to the eye, the
fair forms that have passed there in other days.' (Poe, 1856)

Man-made landscapes can therefore be more beautiful
than natural landscapes, as long as the relationship between
the two contexts, as in every other artificial case, are carefully
designed, as the best architecture of the past centuries has
taught us.

This kind of problem was faced by the architects and
various specialists who, in Malibu, on the Californian coast,
executed the well-known reconstruction of a Roman villa (the
Villa dei Papiri of Herculaneum, buried by the Vesuvius erup-
tion) by Paul Getty; a rather meticulous undertaking accomplished with almost all the same materials as the exemplar;
thus giving rise to a pseudo-replica, which requires a great
deal of maintenance since it is situated a few steps away from
the Pacific Ocean. Moreover, the same difficulties have to be
faced not only by the people who take care of zoos, which we
mentioned in a previous chapter, but also by those who try to
reproduce natural environments in which an animal or vegetable species, such as aquaria, pods, nests, reserves, etc., can
survive.

10 Virtual Reality

Deception, in the broader sense intended here, illusion, a
whole series of side effects, but also the potential utility of the
artificial, converge, in the end, in Virtual Reality technology.
This technology consists of devices which generate three-
dimensional moving pictures on a stereoscopic monitor applied

in front of the eyes in a special helmet. It is an artificial environment in which it is possible to interact, for example, by moving in a room or exploring the human body from the inside. Although this technology has aroused the usual controversy between enthusiastic supporters and detractors who are afraid of the possible 'loss of a sense of identity and reality', some of the most interesting applications are once again in the field of medicine. Indeed, it is possible, by means of virtual reality machines, to visit a patient or operate (telesurgery) on him even though he is far away from the doctor. In 1995, the demonstration presented by the ARPA (Advanced Research Projects Agency) biomedical program was oriented in this direction; this program consisted in carrying out a surgical procedure with a robot which acted on the command of doctors who 'operated' using a monitor many kilometers away from the patient. Thus, we have a chain of artificial objects (the vision of the patient's body by means of telecameras and computers, the robot's arms and hands which intervene on the physical reality of the patient) whose coordination presents several difficulties, including the transmission speed of the signals in both directions which is perhaps the most important.

The most curious aspect – which calls to mind man's ability to adapt to the machines he uses – consists in the possibility that the surgeon, adapting to the delay which is generated between the moment in which he 'acts' on the monitor and the moment in which the robot acts on the patient, may become confused when he takes off his helmet. Moreover, according to experiments conducted by the British Defence Research Agency, virtual reality devices can generate several psycho-physical problems: out of 146 adults in perfect health who used the helmet for 20 minutes, 89 suffered temporary nausea, dizziness or eyesight trouble and 8 did not even succeed in terminating the experiment (Langreth, 1994).

At any rate, in order to provide an effective aid to surgery, many believe that the essential performance chosen up to this point, namely the movement and thus the action of the surgeon's hands, is not sufficient. In fact, these critic believe that it is necessary to introduce devices into the system which are able to reproduce other typical elements of a surgical operation, such as odors and the physical sensations which are associated with the forces involved. Additional research is dealing with these problems as well, and even in this field we can foresee the problem of the model on the basis of which the various performances will be coordinated. According to a report by the American National Institute of Standards and Technology, if

'The manipulation of instruments by the surgeon or assistants can be direct or via virtual environments. In the latter case, a robot reproduces the movements of humans using virtual instruments. The precision of the operation may be augmented by data/images superimposed on the virtual patient. In this manner the surgeon's abilities are enhanced' (Moline, 2001)

but

'the sense of smell in virtual environment systems has been largely ignored. Both Krueger and Keller are developing odor-sensing systems. Smells are extremely important. Not only do they help distinguish specific substances, but also they give a sense of reality to a situation. The absence of odor is a serious limitation of current telepresence and training systems. Another major research problem relates to overlaying ultrasound images on live video that is then viewed in a head-mounted device application. The research issue to be addressed is the alignment of images in real time' (ibidem).

Anyway, according to the report,

'Today simulations trade off less realism for more real-time interactivity because of limited computing power, but the future holds promise of a virtual cadaver nearly indistinguishable from a real person.' (ibidem).

Virtuality is not a prerogative of advanced technology, since our imagination is its first creator and this is why man has always created objects or machines capable of stimulating it, with or without a computer.

Environments and landscapes can also be observed through a window, as the American company Bio-Brite, from Maryland, has intuited.

'People prefer rooms with windows, and for a good reason. Specific research has demonstrated that windows can raise people's morale and even increase productivity. Now it is possible to take advantage of Window-Lite, a pleasant and economic solution for offices, apartments, hospitals, workplaces or other places which are not equipped with windows' (Bio-brite, 2000).

It is a sort of illuminated picture which reproduces the structure of a window and allows one to admire landscapes which range from classical Hawaii to tropical beaches, while creating the sensation of space and reducing the symptoms of claustrophobia. All things considered, we find ourselves before a technological remake of the quadraturism and the *trompe l'oeil* which date back to the XVI century, though with a more direct intention of obtaining realistic effects thanks to the technology available today. A technology which allows the same above-mentioned company to offer a Sunrise Clock Dawn Simulator piloted by an electronic watch and other devices, which acts as an alarm clock, but

'(...) by reversing the negative effects of a gloomy winter day, minimizing various physiological problems associated with daily rhythms and helping people to wake up in a natural way' (Bio-Brite, ibidem).

It is interesting to note how these same topics are more and more commonly the object of study with regard to music and its effects on the human or animal organism. As we have already mentioned, this is not the most appropriate place to discuss the various forms of art as activities that reproduce reality which are thus analogically interpretable according to the artificial theory. Nevertheless, we must emphasize how art itself – for example painting and music, before and after the invention of artificial perspective and notation – has always been devoted to reproducing natural phenomena, even though it generally has poetic rather than 'deceptive' aims.

Apart from the almost obvious case of figurative painting, the work of musical composers is perhaps more significant. Specifically regarding the seasons of the year and other natural events, they have often tried to reproduce their most striking essential performance according to their own poetics. At the same time, they give the public ample opportunity to identify the exemplar and its performances thanks to the expert use of instrumental technology.

Among the most important examples we find the *Four Seasons* by A. Vivaldi, the twelve pieces *The seasons*, for piano, by P.I. Ciaikovsky and then *The Moldavian* by B. Smetana, the *Musical portrait of nature* by J.H. Knecht, the concertos *The hunt, The night, The goldfinch* by Vivaldi, *The sea-storm* by I.J. Holzbauer, the famous pieces for harpsichord by F. Couperin *Les papillons, Le moucheron, Le rossignol en amour*, and the pieces by P. Rameau *Le rappel des oiseaux, La poule*. Other noteworthy examples include the N. 6 Symphony by L. van Beethoven Pastoral, *The fountains of Rome* and *The Pine trees of Rome* by O. Respighi, who also wrote *The birds*, a suite for small orchestra, and the compositions by O. Messiaen *Réveil des oiseaux* and *Oiseaux exotiques*. The skillful presentation of naturalistic effects at the beginning of *Peter and the wolf* by S. Prokofiev is also famous.

It is pointless to add that in all of these works, and in the thousand others of this kind, the descriptive and reproductive aspect is accompanied by 'side effects' which are not at all secondary. This is probably why Beethoven recommended that music must never be obscured by the 'pictorialness' which composers are able to generate.

11 Conclusions

The concept of the artificial together with its related technology, has always been a central part of human activity. With its own characteristics which are clearly distinguishable from those of other activities.

Despite the generic use which we commonly make of this term, almost as if it were interchangeable with the adjective 'technological' or with the expression 'man-made', the theory of the artificial shows how designing in order to produce *ex novo* and designing in order to reproduce something that exists in nature are two distinctly different activities, requiring different abilities and likewise showing different strengths, difficulties and limits. These two activities, which we have called conventional technology and technology of the artificial, respectively generate processes or machines as non-natural realities. In other words, while conventional technology aims right from the beginning to give rise to things which do not exist in nature, the technology of the artificial, though it wishes to generate things inspired by natural exemplars, cannot avoid transfiguring them to some degree.

A world made up exclusively of conventional technological objects and machines would be a world in which two realities, the natural and the technological, would be, as in fact they are, very distinguishable from each other just as a city can be distinguished from the surrounding countryside or, better yet, a car 'cemetery' from the landscape in which it is situated, or a watch or bracelet from human being's wrist or a ball-point pen from his fingers.

Conversely, the artificial, at least ideally, aims to create realities which at certain observation levels, should not be distinguishable from the natural context, whether it be the land, the human body or any other natural environment. The artificial, in fact, stems from an ancient desire not only to control nature but to reproduce it using different strategies than its own.

All this is true in motivational terms, though, in teleological terms, even the artificial could be understood as a matter of a form of control, but of a higher and more ambitious kind, since one expects to substitute the highest possible form of command, that is, the capability to create nature *ex novo*.

Nevertheless, as we have said, in the end both technologies create products that are recombinations of natural elements. In the case of the artificial, we face many difficulties in reaching the aim of reproducing the natural exemplars and, moreover, we get effects which are different from the desired targets.

Perhaps it is the revenge of natural reality, but the fact is that conventional technology, with its materials and techniques, poses serious limitations on the technology of the artificial. In fact, on one hand artificial technology allows, at certain observation levels, to 'deceive' an organism, a spectator, an environment, a living species or other structures or natural events. On the other hand, it reproposes, at different levels and often also at the one assumed by the designer, its own heterogeneity compared to the exemplar and its performances. Hence, in spite of the artificialists' aims, the artificial tends to create realities which deviate more and more from nature, just as conventional technology does deliberately right from the beginning.

This does not mean that the artificial is intrinsically destined to fail: in fact, we know that it has had some success. However, in producing the artificial, one has to keep in mind that the similarity with what happens in the natural exemplar, will always be accompanied by many unavoidable effects or properties whose interplay with the selected essential performance may generate a 'new version' of the natural phenomenon.

Thus, we can maintain that artificialism has its own unique philosophy which differentiates it from conventional

technology, above all in terms of ideation, design and constraints. In particular, we have established the following
points.

- The artificial always derives from a process of multiple
  choices which, as such, prevents it from pursuing the
  overall analytical reproduction of the natural object or
  process which it intends to reproduce. These choices
  pertain to observation level, exemplar and essential
  performance.

- The selection of an observation level, which also includes the possibility of its deliberate construction
  without any direct reference to the sensible world,
  leads to the formation of individual or collective representations or models – of the natural object observed –
  which do not necessarily respect 'reality as it is'. Representations may depend on a whole series of cultural
  premises, principles, preferences, beliefs or prohibi-
  tions which history is full of.

- The selection of the exemplar, in turn, inevitably assumes the form of an isolation, and at times a real
  eradication, of the object or process from the natural
  context where it is found. In some cases, the resulting
  model does not have serious consequences for the reproducibility of the exemplar at the chosen observation
  level. In many other cases, however, its isolation cuts
  off, so to speak, relationships at various observation
  levels with the rest of the context which it belongs to.
  These relationships could be and often are vitally important in determining the characteristics of the natural object or process and, therefore, the resulting artificial object or process.

- The selection of the essential performance is already, on the one hand, bound by the selection of the observation level and exemplar and, on the other hand, almost always constitutes a kind of 'bet' on the quality, function or based which is considered 'fundamental' in the exemplar. Furthermore, the selection of the essential performance, implies another process of isolation of that performance from the other performances or properties which characterize the exemplar's way of being in nature. In other words, the selected essential performance is, almost always, unavoidably modellized as a 'purified' performance, i. e. as if it were a 'context free' property or based.

- The principle of 'functional equivalence' – between the exemplar and its formal model or its concrete reproduction – only guarantees that, in a few circumstances, the essential performance is sufficiently autonomous and therefore transferable to different supports, but does not guarantee that such supports reproduce the structure of the exemplars. Hence, there is no reason for the properties of an artificial device – apart from the selected and reproduced essential performance – to overlap the properties of the natural exemplar.

- The emergence of properties, qualities and performances which are not explicitly planned in the design is thus inevitable for the artificial. In any case, this does not necessarily mean that whatever emerges from an artificial object or process will correspond with what we would expect to emerge from the natural exemplar. It is unlikely that the whole network of the relationships among all the observation levels present in the will overlap those of the artificial, because what is ex-

pected and implicit in the model which guides the reproduction is always a reality considered at only one
level, that is, from only one perspective among the infinite possibilities.

- Compatibility with the host context, environment or
  organism is usually one of the most crucial moments
  for the artificial, since it is precisely at that stage that
  the heterogeneity of its structure or its performances
  will clearly emerge in terms of 'boundary problems'.

Deceit, simulation and illusion can certainly be usefully
pursued using various strategies, but the inevitably rigid limits
within which they are possible, rely heavily on the essential
performance's autonomy from the 'support' which generates it
in nature. At times this autonomy can be relatively high as it is
when we are dealing with exemplars or, even more so, informational performances, while it is usually very low in all the other
concrete cases.

- The synthesis or coordination of more than one artificial device (each based on its own exemplar and essential performance) creates further problems. Indeed, the
  'sum' of two artificial objects or processes only in-
  creases the distance of the presumable resulting object
  or process from the system made up of the two exemplars in nature. This is mainly due to the relationships, within the exemplar, among structures and
  processes at observation levels which are inevitably
  neglected in the model which guides the reproduction.
  This could be called the 'unavoidable price of analysis'.

The design of a *third level coordinator* of two artificial devices can be conceived as an expedient or as a new artificialistic target. While in the former case the resulting object or

process will have no intended similarity with what is found in nature, in the latter case we are dealing with quite a new undertaking. Indeed, the reproduction of the coordinator could require giving up or changing the observation levels assumed for designing the first two artificial objects or processes. In turn, this could involve changes in the structure or the performances of the two artificial objects or processes involved, so that they fit the requirements of the third observation level. Thus, a sort of *ad infinitum* rebound between bottom-up and top-down strategies could take place.

In any case, though today the synthesis of two or more artificial subsystems of a natural system is almost impossible if one expects the same performances we get from the natural exemplar (above all the performances that were not explicitly designed) from that synthesis, we should not overlook the fact that *something real always happens* in such attempts.

- The transfiguration of the exemplar and its performances, and often the essential performance as well, constitutes an inexorable tendency of artificial objects and processes. It is due to the combined effect of the selective and heterogeneity factors mentioned above. Since nature, including human nature, is a whole which is intrinsically integrated – by physical, chemical, biological, psychological, and sociological laws – even the optimal reproduction of one performance of an exemplar already constitutes an anomaly in itself, because it does not have, nor can it easily have, natural relationships with the host context at all the natural levels involved when we consider it in nature.

- The final and best result of an artificial device or process, may be to locally exhibit an acceptable ability to reproduce the selected essential performance of a natural exemplar, under a quantitatively acceptable set of operational conditions, accompanied by the lowest quantity of side effects.

No matter what actually happens and will happen through the research which is planned in many fields of the artificial, we can be certain that from whatever perspective it is examined, the artificial tends to generate a separate reality, which should be studied in depth, and will be in the coming years. Indeed, the density of artificial objects, processes and machines is increasing steadily, thanks to more refined technologies which allow man to renew his utopic visions, and because many aspects of our existence have always been ready and available to gain practical and imaginative, scientific or artistic benefits from those technologies.

It is likely that the dream of realizing a cyborg (cybernetic organism, that is, a natural man who is improved and amplified in his performances by artificial devices connected directly to his body) will remain unattainable for a long time, if we take the exaggerated characters of science fiction films as examples. Nevertheless, as we have seen, today millions of people live, work, communicate or move thanks to extensions, prostheses or implants of artificial objects of various kinds. The prospect of a more marked physical integration with artificial is therefore not unthinkable. On the contrary, serious sociological, psychological and even ethical problems are already appearing on the scene. We do not know exactly what this will entail. However, to quote a very profound remark by Willelm Kolff, it is possible that in one of the next Olympic games the marathon will be won by a man equipped with an artificial heart, which is stronger and more resistant to fatigue than the natural heart. The crucial point is this: will he be disqualified?

Some Uses of the Word Artificial

Premise

Generally, the many kinds of artificial devices normally defined as automatisms are excluded. Moreover, some of the objects or processes which are defined here as artificial do not correspond with the definition of artificial introduced in this book. On the contrary, they constitute the result of a patchwork or recombination of the same materials used in their exemplars.

Before 1800

artificial classification
    (Linneo)
artificial fireworks
artificial flower
artificial ice
artificial insemination
    (L. Spallanzani)
artificial irrigation
artificial island
artificial lake

artificial marble
artificial propagation (agri-
    cultural botany)
artificial rainbow (F. Bacone)
artificial writing
    (J. Gutenberg)
memoria artificiosa (Cosmas
    Rossellius)
perspectiva artificialis
    (L.B. Alberti, P. della
    Francesca)

After 1800

artificial adaptation (J.H. Holland)
artificial arm (V. Kolff)
artificial bait
artificial barrier
artificial beach
artificial bells
artificial blood (R. Naito)
artificial blue (J. Guimet)
artificial bone
artificial breeding
artificial cavity
artificial cell (T.M.S. Chang)
artificial chemical element
artificial chromosome
artificial colours (A. Baeyer)
artificial cornea
artificial diamond (P.Williams Bridgman)
artificial ear
artificial environment
artificial esophagus
artificial experts (M.H. Collins)
artificial extremities
artificial eye
artificial fertilizers (J.B. Lawes)
artificial fibres

artificial fish (M.S. Triantafyllou)
artificial flavour
artificial flower
artificial gelatin
artificial grass
artificial gravity
artificial ground
artificial habitat
artificial hair
artificial hatching
artificial heart (W. Kolff, D. Liotta)
artificial hip
artificial honey
artificial horizon
artificial incubation
artificial insemination
artificial intelligence (M. Minsky, H. Simon et al.)
artificial island
artificial ivory (H. Scarton and Calabrese)
artificial joints
artificial kidney (W. Kolff)
artificial lake
artificial landscape
artificial language
artificial larynx

artificial leather
artificial leaves (S. Winkler)
artificial life (C.G. Langton)
artificial ligament
artificial light (T. Edison)
artificial limbs (various)
artificial liver (A. Demetriou)
artificial lung (G. Mortensen)
artificial milk
artificial muscle
artificial nail
artificial nest
artificial organs (various)
artificial pancreas
artificial paradises (C. Baudelaire)
artificial pearl (M. Koukichi)
artificial perfume
artificial plant
artificial pond
artificial protein
artificial radioactivity (F. Joliot)
artificial rain (I. Langmuir, V. Schaefer)
artificial reality (M. Krueger)
artificial reef

artificial resin
artificial respiration
artificial retina (M.A. Mahowald, C. Mead)
artificial rock
artificial rubber (W. Carothers, J. Nieuwland)
artificial satellite
artificial sensation
artificial silk (H. de Chardonnet)
artificial skeleton
artificial skin (J. Burke, I. Yannas)
artificial smell (Sony Corp.)
artificial snow (Emile Wyss & Cie SA)
artificial sound
artificial speech
artificial star
artificial sweetener
artificial taste
artificial tears
artificial vessels
artificial virus
artificial weathering

Examples of conventional technology

1602 Medical thermometer (Santorio, Galileo)

1608 Telescope (Lippershey)

1609 Printed newspaper (Aviso – Relation oder Zeitung)

1643 Barometer (Torricelli)

1657 Pendulum clock (Huygens)

1682 Pressure cooker (Papin)

1698 Steam engine (Savery)

1733 Device for the sewing machine (Kay)

1745 Electrical capacitor (Kleist and Van Musschenbroek)

1769 Motor vehicle (Cugnot)

1775 Streetcar (Outram)

1781 Centrifugal governor (Watt)

1800 Electrical cell (Volta)

1806 Carbon paper (Wedgwood)

1817 Bicycle (Von Drais)

1818 Stethoscope (Laennec)

1818 Steering wheel (Ackermann)

1829 Piano (Demian)

1831 Electrical engine (Henry)

1835 Relay (Henry)

1835 Revolver (Colt)

1837 Ship propeller (Ericsson)

1843 Fax (Bain)

1844 Process of vulcanization of rubber (Goodyear)

1844 Telegraph (Morse)

1846 Nitroglycerin (Sobrero)

1846 Saxophone (Sax)

1851 Gyroscope (Foucault)

1857 Elevator (Otis)

1859 Storage battery (Plant)

1860 Linoleum (Walton)

1861 Rotary press (Hoe)

1862 Machine gun (Gatling)

1865 Generator (Pacinotti)

1866 Dynamite (Nobel)

1868 Celluloid (Hyatt)

1869 Compressed air brakes (Westinghouse)

1869   Chewing gum (Semple)
1871   Drill (Morrisson)
1871   Telephone (Meucci)
1877   Gramophone recorder (Cros)
1879   Electric lamp (Edison)
1879   Cash register (Ritty)
1880   Cable railway (Olivieri)
1882   Power station (Edison)
1883   Fountain-pen (Waterman)
1884   Steam turbine (Parsons)
1886   Linotype (Merenthaler)
1886   Aluminum (Hall and Hiroult)
1888   Straw (Stone)
1888   Tire (Dunlop)
1889   Eiffel Tower (Eiffel)
1890   Electric chair (Brown and Kennally)
1890   Electromagnetic wave detector (Branly)
1892   Thermos (Dewar)
1892   Reinforced concrete (Hennebique)
1892   Escalator (Reno)
1893   Zipper (Judson)
1894   Sphygmomanometer (Riva Rocci)

1895   Slot machine (Fey)
1899   Magnetic recorder (Poulsen)
1902   Disk brakes (Lanchester)
1905   Diode (Fleming)
1907   Joystick (Esnault-Pelterie)
1908   Vacuum cleaner (Spengler)
1908   Cellophane (Edwin)
1909   Neon tube (Claude)
1909   Toxic fume suppresser (Frenkel)
1910   Radioactivity counter (Geiger)
1913   Petroleum pump (...)
1914   Stoplight (...)
1914   Brassière (Jacob)
1915   Pyrex pot (Litteton)
1915   Tank (Swinton)
1920   Mass spectrograph (Aston)
1921   Cement (Dickson)
1924   Electroencephalograph (Berger)
1926   Television (Baird)
1926   Combustible fluid for missiles (Goddard)
1927   Radio compass (Busignies)
1930   Helicopter (D'Ascanio)
1934   Nylon (Carothers)
1935   Lie detector (Keeler)

1938   Instant coffee
(Nestlè)
1938   Photocopier
(Carlson)
1939   Spray (Kahn)
1939   Jet (Von Ohain)
1942   Atomic battery
(Fermi)
1943   Ball-point pen (Biro)
1945   Atomic bomb
(Oppenheimer)
1946   Microwave oven
(Spencer)
1946   Radial tires
(Michelin)
1947   Pinball machine
(Mabs)
1948   Transistor
(Shockley, Brattain,
Bardeen)
1951   Tetra-pack (Reusing)

1954   Solar battery
(Pearson)
1955   Optical fibre
(Kapany)
1958   Integrated circuit
(Kilby)
1959   Hovercraft
(Cockerell)
1960   Laser (Maiman)
1963   Tape cassette
(Philips)
1965   Crystal liquid monitor (Heilmeier)
1971   Microprocessor
(Faggin, Hoff, Mazer)
1972   TAC (Hounsfield)
1978   Modem (Hayes)
1978   Floppy disk (Apple,
Tandy)
1982   Compact disk
(Philips, Sony)

Bibliography*

AFCEA International Press, 1988, p. 60.

R. ARNHEIM, *Entropy And Art. An Essay On Disorder And Order,* University of California, (1971), It Transl. *Entropia e arte,* Einaudi, Torino 1974, p. 53.

D. BAGGI, (ed), *Readings in Computer Generated Music,* IEEE, Los Alamitos, Ca 1992.

D. BALASUBRAMANIAN, 'New eyes for old?, The Hindu, Adapted from the Convocation Address given at the Elite School of Optometry', Chennai, on September 5, 1998, Internet site: http://www.webpage.com 1998.

S. A. BEDINI, 'The role of automata in the history of technology', Technology and Culture, 5, 1, 1964.

T. W. BERGER, in E. MANKIN, *Plugging into the brain,* University of Suthern California Chronicle, Los Angeles 1995.

D. BERTASIO, *Studi di sociologia dell'arte,* Franco Angeli, Milano 1996.

BIO-BRITE, Internet site: http://www.biobrite.com 2000.

BIO-VASCULAR, 'Movements on-line', Internet site: http://www.ioi.com.

BISCHOFF, V. GRAEFE, K. P. WERSHOFEN, 'Combining object-oriented vision and behavior-based robot control', Report, Institute of Measurement Science, Federal Armed Forces University Munich 1996.

F. BOAS, *Primitive Art,* Instituttet for sammenlignende kulturforskning, Oslo 1927

H. BREDEKAMP, *Antikensehnsucht und Maschinenglauben,* Klaus Wagenbach, Berlin 1993, tr. it. *Nostalgia dell'antico e fascino della macchina,* Il Saggiatore, Milano 1996, p.112.

S. BUTLER, *Erewhon,* Adelphi, Milano 1988. Electronic Text Center, University of Virginia Library, p. 250.

---

* For Internet addresses the character ⏎ indicates a line break and should not be typed as part of the address. Please note that some addresses may be incorrect, since they are frequently updated or removed.

G. P. CESERANI, *I falsi Adami*, Milano 1969, p. 88.

T. L. CHANG, 'Interview' by Anfré Picard for McGill News, Alumni Quarterly, Internet site: http://www.mcgill.ca 1996.

C. W. CHURCHMAN, 'The artificiality of science: Riview of Herbert A. Simon's book *The Sciences of the Artificial*', Contemporary Psychology, 15, 6, June 1979, p. 385-386.

H. M. COLLINS, *Artificial Experts: Social Knowledge and Intelligent Machines,* The MIT Press, Cambridge, Mass. 1990.

J. E. COOKE, Internet site: www.pharmacy.ualberta.ca 2000.

R. CORDESCHI, 'The discovery of the artificial. Some protocybernetic developments 1930 – 1940', Artificial intelligence & Society, London 1991.

B. CRANDALL, B. LEWIS, (Eds) *Nanotechnology: Research Perspectives*, The MIT Press, Cambridge, Ma 1997, pp.vii-viii.

H. de MONANTHEUIL, in *Quaestiones mechanicae* 1517.

M. DENIS, *Image and cognition*, Presses Universitaires de France, Paris 1989.

K. E. DREXLER, *Engines of Creations: Challenges and Choices of the Last Technological Revolution*, Doubleday, New York 1986, p. 14.

FLAVOR CONCEPTS, Inc, Internet site: http://flavorconcepts.com.

D. M. FRYER, J. C. MARSHALL, 'The motives of Jacques de Vaucanson', Technology and Culture, 20, 1979, pp. 257-269.

M. GALLONI, 'Microscopi e microscopie, dalle origini al XIX secolo', Quaderni di storia della tecnologia, Levrotto & Bella, 3, 1993, p. 23.

W. W. GIBBS, Artificial Muscle, Internet site: http://cape.uwaterloo.ca 1998.

A. GIDDENS, 'Structuralism, post-structuralism, and the production of culture', in Anthony Giddens and Ralph Turner (eds.), *Social Theory Today*, University of Stanford Press, Stanford, CA. 1987.

C. GOLDMAN, 'Artificial Lakes Face Real Environmental Conflict', excerpted from The Davis Enterprise Wednesday, Oct. 6 1999.

S. HARNAD, 'Levels of Functional Equivalence in Reverse Bioengineering. The Darwinian Turing Test for Artificial Life', Artificial Life 1(3) 1994.

S. HAYKIN, *Neural Networks: A Comprehensive Foundation*, Macmillan, New York 1994, p. 2.

HITACHI INSTRUMENTS, Inc., Hitachi Diffraction Gratings, Internet site: http://www.hii.hitachi.com.

W. HOFFMAN, 'Forging new Bonds', in *Inventing Tomorrow*, University of Minnesota Institute of Technology, Spring 1995.

R. HOFFMANN, S. LEIBOWITZ SCHMIDT, *Old Wine New Flasks*, Freeman and Company, New York 1997, p. 19-20.

DIP-INFM, University of Genoa, Internet site http://www.sts.tu-harburg.de.

S. A. EDWARDS, Engineering Human Tissue, http://mem⏎bers.aol.com 2001.

M. KAWANISHI, 'Developing a Method to Assess the Barrier Performance of High-Level Radioactive Waste Disposal Facilities', Abiko Research Laboratory, Nuclear Fuel Cycle Project Department, Internet site: http://criepi.denken.or.jp 1995.

J. F. KÖHLER, 'Historia Scholarum Lipsiensium' 1776, quoted in H. T. David, and A. Mendel, *The Bach Reader*, Dent, London 1946, seen in J. Amis, M. Rose, *Words about Music*, Faber and Faber, London 1989, p. 186.

T. KUHN, *The Structure of Scientific Revolutions*, Chicago University Press, Chicago 1970.

R. KURZWEIL, *The Age of Spiritual Machines. When Computers Exceed Human Intelligence*, Penguin Books New York 2000.

T. JEFFERSON, Thomas Jefferson to John Adams, October 28, 1813, Internet site: http://128.143.230.66.

P. N. JOHNSON-LAIRD, *Mental Models. Towards a Cognitive Science of Language*, Cambridge Universiy Press, Cambridge, p. 45-46.

B. JOY, 'Why the future does not need us', WIRED, 8, April 2000.

T. KEAVENY, 'Presentation page for the Berkeley Orthopaedic Biomechanics Research', Web Internet site, http://bio↵ mech2.me.berkeley.edu 1996.

R. J. KERN, 'Artificial Reefs: Trash to Treasure', National Geographic News, Internet site: http://news.national↵ geographic.com.

G. P. LANDOW, *Hypertext: The Convergence of Contemporary Critical Theory and Technology*, Johns Hopkins University Press, London 1992.

R. LANGRETH, 'Virtual Reality: Head Mounted Distress', Popular Science, 245, 5, 1994 p. 49.

C. G. LANGTON, 'Artificial Life', in C. G. Langton, (ed.) *Artificial Life*, Santa Fe Institute Studies in the Science of Complexity, Addison-Wesley, Redwood City, Ca. 1989, p. 32.

C. G. LANGTON, 'Preface.' In C. G. Langton, C. Taylor, J. D. Farmer, and S. Rasmussen, editors, Artificial Life II, Volume X of SFI Studies in the Sciences of Complexity, Addison-Wesley, Redwood City, CA 1992, pages xiii-xviii.

M. LEE, A. FREED, D. WESSEL, 'Real-time Network Processing of Gestural and Acoustic Signals', CNMAT, University of California 1991, internal report.

D. LEISER, G. CELLERIER, J. J. DUCRET, 'Une ètude de la fonction representative', Archive de Psychologie, XLIV 1976, p. 171.

M. LESCHIUTTA, S. ROLANDO 'I primi strumenti di misura e-lettrici,' Quaderni di storia della tecnologia Vol 3 1993, pp. 58-59.

M. LOSANO, *Storie di automi*, Einaudi, Torino 1990, p. 128.

Larson Utility Camouflage, (2001), Internet site http://www.utilitycamo.com.

M. A. MAHOWALD, C. MEAD, 'The silicon retina', Le Scienze-Scientific American 1991, p. 275.

G. R. MARTIN, 'The visual problems of nocturnal migration', Bird Migration, Springer-Verlag, Berlin 1990, pp. 185-197.

B. MAZLISH, 'The man-machine and artificial intelligence', SEHR, volume 4, issue 2: Constructions of the Mind.

McGOWAN Center, Internet site http://www.upmc.edu.

A. G. MIKOS, R. BIZIOS, K. K. WU, M. J. YASZEMSKI, 'Cell Transplantation', The Rice Institute of Biosciences and Bio-engineering, Web site in Internet http://www.bioc.rice.edu 1996.

M. MINSKY, 'Will Robots Inherit the Earth?', Scientific American, Oct, 1994.

J. MOLINE, 'Virtual Reality for Health Care: a survey', National Institute of Standards and Technology, Gaithersburg, Maryland 20899.

J. MONGE-NÁJERA, M. BLANCO, 'Plants that live on leaves', Internet site, http://www.ots.duke.edu 2001.

H. MORAVEC, 'The Robot as Liberation from Human Nature', transcript of 1989 Hull Memorial Lecture, in INTERACTIONS: 10, Worcester Polytechnic Institute, Worcester, Massachusetts, December 1989, pp. 32-42.

NATIONAL INSTITUTE OF HEALTH, 'Clinical Applications of Biomaterials. NIH Consens Statement', Nov 1-3, 1982, 4(5).

J. NEEDHAM, *Science and Civilization in China*, C ambridge University Press, Cambridge 1975, p.53.

N. NEGROPONTE, *Essere digitali*, Sperling & Kupfer Milano 1995, p.123.

M. NEGROTTI (ed.), *Understanding the Artificial*, Springer Verlag, London 1991, tr. it., *Capire l'artificiale*, Bollati-Boringhieri, Torino 1990, 1993.

E. A. POE, 'The Landscape Garden' (C), The Works of the Late Edgar Allan Poe, vol. IV, Redfield, New York 1856, pp. 336-345.

H. POINCARE', 'Mathematical Creation', in B. Ghiselin (ed.), *The Creative Process*, Univ. of California Press, Los Angeles 1952, p.33.

M. POLANYI, *The Tacit Dimension*, Doubleday, New York 1966.

R. POSNER, 'What is culture? Toward a semiotic explication of anthropological concepts', in W. A. Koch (ed.), *The Nature of Culture*, Brockmeyer, Bochum 1989.

L. QVORTRUP, 'Sistemi naturali, sociali e artificiali: verso una tassonomia dell'artificiale', in NEGROTTI M. (ed.), *Artificialia*, Clueb, Bologna 1995.

T. REGGE, *Infinito, viaggio ai limiti dell'universo*, Mondadori, Milano 1994, p.142.

SAN JOSE MERCURY-NEWS: CONSUMER CORNER, 1997, Internet site: http://www.gorp.com.

M. SCHNEIDER, *Le rôle de la musique dans la mythologie et les rites des civilisations non europénnes*, éditions Gallimard, Paris 1960, tr. it., *La musica primitiva*, Adelphi Edizioni, Milano 1992, p.35.

M. RIZZOTTI, M. 'Il concetto di artificiale', Memorie dell'Istituto Veneto di Scienze, Lettere ed Arti, Venezia 1984.

R. ROSEN, 'Bionics Revisited', in H. HAKEN, A. KARLQVIST, U. SVEDIN, (eds.), *The Machine as Metaphor and Tool*, Springer Verlag, Berlin Heidelberg 1993.

K. SANES, Internet site: http://www.transparencynow.com 1998.

J. SEARLE, *Minds, Brains and Science*, Harvard University Press, Cambridge, Mass. 1984.

H. A. SIMON, *The Sciences of the Artificial*, MIT Press, Cambridge Mass. 1969, tr. it. *Le scienze dell'artificiale*, ISEDI, Milano 1970.

J. A. SLOBODA, *The Musical Mind. The Cognitive Psychology of Music*, Oxford University Press, Oxford 1985.

D. SOLLA PRICE (de), 'Automata and the Origin of Mechanism', TECHNOLOGY AND CULTURE, V, 1 1964, p. 8.

SOMSO MODELS, Internet site: http://www.holtanatomi⌐cal.com.

S. SUSKA, 'Adjusting to Nature in Cape Hatteras National Seashore' 1998, condensed version of Dolan and Bruce Hayden 'Adjusting to Nature in Our National Seashores' by Robert which originally appeared, 'National Parks & Conservation Magazine', June 1974, Internet site: http://www.nps.gov.

M. TABOR, 'Michael Tabor's IGERT' Statement Internet site: http://w3.arizona.edu.

Y. TAKIMOTO 'The Experimental Replacement Of A Cervical Esophageal Segment With An Artificial Prosthesis With The Use Of Collagen Matrix And A Silicone Stent', 6th Internet World Congress for Biomedical Sciences, 2000, Presentation # 177.

G. TARDE, *Les lois de l'imitation*, Paris 1890, tr. it. *Le leggi dell'imitazione*, UTET, Torino 1976.

TEIJIN Teijin Public Relations & Investor Relations Office, Teijin Develops Greener Process for Making Superb Artificial Leather, Internet site: http://www.teijin.co.jp 2000.

TEXAS WATER RESOURCES, 'Replicating Mother Nature', Vol. 18 Number 1, Spring 1992.

M. S. TRIANTAFYLLOU, G. S. TRIANTAFYLLOU, 'Un robot che simula il nuoto dei pesci', 1995, Le Scienze-Scientific American, 3, 1996, p. 321.

F. J. VERHEIJEN, 'The mechanisms of the trapping effect of artificial light sources upon animals', Netherlands Journal of Zoology, 13, 1958, pp. 1-107.

S. VOGEL, *Cats' Paws and Catapults: Mechanical Worlds of Nature and People*, W. W. Norton & Company, Inc, New York 2000.

I. WEINTRAUB, 'Pressure Used to Create Artificial Diamonds', The Physics Factbook Edited by Glenn Elert, Internet site: http://hypertextbook.com 1998.

N. WIENER, *The Care and Feeding of Idea*, MIT Press, Cambridge, Ma. 1993.

R. M. WINSLOW, 'Ask the expert, How do scientists make artificial blood? How effective is it compared with the real thing?', Scientific American, Nov. 2001.

C. WITTENBURG, et al., 'Abstract', Universität Hamburg, Institut für Anorganische und Angewandte Chemie, Internet site: http://www.chemie.uni-hamburg.de 2000.

WHOLESALE NUTRITION, Internet site, http://www.nutri.com 1996.

B. WOOLLEY, *Virtual Worlds. A Journey in Hype and Hyper-reality*, Blackwell, Oxford 1992.

J. Z. YOUNG, *A Model of the Brain*, Oxford University Press, Oxford 1964, tr. it., *Un modello del cervello*, Einaudi, Torino 1974, p. 278.